INTERNATIONAL ENERGY AGENCY

THE ROAD FROM KYOTO

Current CO_2 and Transport Policies in the IEA

FOREWORD

Transport accounts for almost a third of total final energy consumption in IEA countries, and 80% of that amount is in the form of road transport. Cars and trucks will be the principal source of carbon dioxide emissions in the foreseeable future. The central role of transport in economic activity and its pervasive influence on every day life make it complex and politically difficult to change. CO_2 emissions are just one of many policy concerns, such as safety, urban quality, local air pollution, noise and congestion, that transport raises. Government at many levels – local, regional and national – deals with fuel taxation, urban and regional planning, transport infrastructure investments and public transport. These activities need to be co-ordinated to achieve an effective transport policy in relation to oil security and climate change.

"The Road from Kyoto" reviews and offers insights into how governments are grappling with the complexity of transport and CO_2 policy making. It examines the transport sector from the perspective of the emissions reduction commitments made under the 1997 Kyoto Protocol. It identifies the components of transport that affect CO_2 emissions and details policies that contribute to emissions reduction. It presents detailed accounts of the current and future situation in six IEA countries – Denmark, Germany, the Netherlands, Sweden, the United Kingdom, as well as the European Union as a whole and the United States. Policies and measures directly targeting CO_2 reduction, as well as those that have an indirect impact on climate change in these six countries, are reviewed. Finally, the lessons learned in each country are discussed.

Robert Priddle
Executive Director

ACKNOWLEDGEMENTS

This report is part of a series of forthcoming IEA publications addressing domestic policies and measures to reduce CO_2 emissions. A short report focusing on emission reduction opportunities from fuel economy improvement will be available in November 2000, and a full-length book covering a wide range of transportation policies and measures will be published early in 2001.

This publication is a joint effort of two divisions of the Office of Energy Efficiency, Technology and R&D under the direction of Hans-Jørgen Koch: the Energy Efficiency Policy Analysis Division and the Energy Technology Policy Division. This work has been supervised by Carmen Difiglio. The main contributors to this study are Lee Schipper, Céline Marie, Lew Fulton, Michael Landwehr and Roger Gorham (now with the World Bank). They are grateful for discussions with many national experts and for their helpful co-operation in providing and verifying factual information.

Preparation of the manuscript was the responsibility of Céline Marie, with the editing assistance of Bill Maly and Scott Sullivan. The production assistance of Corinne Hayworth and Muriel Custodio added significantly to the material presented.

TABLE OF CONTENTS

INTRODUCTION

Transport in the Framework of the Kyoto Protocol

In December 1997, leaders of the world's governments met in Kyoto, Japan, to discuss a plan for reducing greenhouse gas emissions from anthropogenic sources, particularly carbon emissions from the use of fossil fuels. While expectations and commitments under the Kyoto Protocol varied greatly among countries, all parties recognised the key role that transportation has played in increasing emissions from fossil fuels.

World-wide CO_2 emissions from fuel combustion have increased over the past 30 years and transport's share of these emissions has also increased. From 1971 to 1997, CO_2 emissions from the transport sector nearly doubled in OECD countries, and the sector's share of total emissions climbed from 19.3 percent to 22.7 percent. Within the transport sector, motorised road traffic accounts for the vast majority of CO_2 emissions. In 1997, road transport produced 81 percent of transport-related CO_2 emissions in OECD countries (IEA 1999), of which 60 percent came from automobiles and household light trucks[1]. The effort by IEA countries to control carbon emissions has so far been focused on road traffic in general, and cars and light trucks in particular.

Even before the Kyoto meeting, the IEA had noted and studied the role of transport in rising CO_2 emissions. The last of several reports on the subject was *Transport, Energy and Climate Change*, which appeared in 1997, and gave an overview of efforts by IEA countries to address rising emissions from transportation. It explored the "pragmatic scope for policy making."

The book you are reading carries forward the Agency's earlier work and provides a tighter focus. It explores recent developments in the policies of six Member countries on road traffic. We review policies that affect

1. The term "light trucks" refers to small passenger vans and sport-utility vehicles, as well as ordinary light trucks, such as pickup trucks, that are used almost exclusively for private, not for goods transport purposes.

transport activity, the fuels used and the amount of fuel consumed per unit of activity in different vehicles. These policies include fuel pricing and other fiscal measures, administrative and regulatory measures and other incentives, including voluntary agreements. Not all the policies considered are aimed at CO_2 reduction, but all affect it in one way or another.

The emphasis here is on which national policies are now in place, when and how they were developed, and where countries appear to be heading in the development of new policies. We review policies that address both technologies and behaviour. We look at some of the most recent adjustments to policies, both those that were underway before the Kyoto meeting took place and those that have been adopted in response to the Kyoto Protocol: one hope is that, by exploring transport policy-making in IEA Member countries, we can learn which policies appear most practical, which are the most difficult to implement, and how experience varies across countries. We can also see what has been accomplished and what work remains for potential future policies.

Future volumes in this series will further explore policies and measures for reducing CO_2 emissions in transport. A short report on fuel economy improvement is being published concurrently with this report.

Linking Energy, Transport, and Environment

Reviewing policies and measures that reduce carbon emissions *directly* is a straightforward affair. More complicated, but equally important, is the review of policies and measures that work *indirectly* – mainly through transport and environmental authorities. We attempt to cover both types of policy in this book. Some of these indirect components may be unfamiliar to the energy policy community, but they have become central to national strategies. They typically vary from country to country. Some policies are more or less practical depending on a given country's particular situation. Consideration of what makes each country unique, and what is common across countries, is also to be found here.

CO_2 emissions reduction is clearly high on the agenda of energy ministries in IEA countries. None of them starts with a clean slate. In every IEA country there are already myriad policies and measures in place that affect CO_2 emissions. For this reason, it is important to look *beyond* CO_2 policies, narrowly defined, and to include policies that affect other spheres of interest and influence.

The Purpose of this Book

This book has three aims: to review existing and emerging polices to reduce CO_2 emissions from transport in the six covered countries (plus the European Union); to describe the history and context in which these policies have been developed; and to consider lessons learned in each country and how these lessons might apply to other countries.

Chapter 1 sets the stage by reviewing general trends in CO_2 emissions in the studied countries as well as world-wide. Chapter 2 provides a framework for considering the impact of transport-relevant policies and categorising these policies by their intended areas of impact. Chapters 3 to 9 present case studies of CO_2-relevant policy-making in six IEA countries, and at EU level. The country case studies discuss what has occurred in each country and explore why different types of policies and measures have been successful in different countries. While this review of policies is mainly qualitative, some numerical examples are also presented. To the extent that there are clear connections between specific policies and CO_2 reductions, these connections are also discussed, although no systematic quantitative analysis is presented that attempts to create such links. This report does not seek to evaluate every single policy with a potential CO_2 connection. Rather, it attempts to cover the *major* policies, and to include a number of transport policies that are *not* primarily intended as CO_2-reduction policies but that may in fact significantly influence CO_2 emissions. Chapter 10 presents the study's conclusions, including comparisons and contrasts of country policy efforts, as well as lessons to be learned from this exercise.

While this book discusses policies to encourage the adoption of new technologies, it does not provide quantitative analysis of specific technologies. Other recent and on-going IEA studies quantitatively examine the potential for transport technology and non-technology policies to reduce CO_2 emissions (IEA 1999 is an example).

Box 1. Overview of the Case Studies

The six countries analysed in this report were selected in part because of their contrasting situations. They vary widely in size and geography, fuel prices, policy approaches, the presence or absence of a domestic car industry, and so forth. The selection features two major vehicle-producing nations (the United States and Germany), one modest producer (the United Kingdom), one small producer (Sweden), and two non-producers (the Netherlands and Denmark). Three of the six countries have high traffic congestion (the United States, the United Kingdom, and Germany) while the others have modest congestion. Per-capita income in the six countries varies by a factor of nearly two, fuel prices by a factor of three, and the tax burden on an average new car by a factor of more than 10.

Background on Each Country. Published studies of the CO_2 reduction potential of each country, and of potential policies, including those put forward by national transportation, environmental, and energy authorities, are reviewed. Published bills and other government policy documents going back to at least 1990 are reviewed, as are academic papers, newspaper articles, published surveys, and other informal sources of information.

Representatives of energy, transport and environment ministries were interviewed, as were a variety of environment activists, academic figures and energy consultants. Several roundtable discussions were held to explore the policy environment in different countries. Each case study was submitted for review by relevant authorities in each country. Discussions were also held with various stakeholders, especially automobile companies – including Renault, Nissan, Honda, Daimler-Benz, Volkswagen, Volvo, General Motors, Ford, and Chrysler – and with the Japan Auto Research Institute.

Timing of Policies Described. All country chapters and conclusions reflect the status of each country's policies as of spring 2000.

Quantitative Analysis of Transport and Emissions Trends. The background trends analysis was carried out and previously reported in a series of IEA publications in collaboration with the Lawrence Berkeley National Laboratory. Key results appeared in two IEA books on energy indicators that were published in 1997.

CHAPTER 1: THE CO_2 PROBLEM — THE POLICY IMPERATIVE AFTER KYOTO

Trends in CO_2 Emissions from Transport

Over the past three decades, carbon dioxide emissions resulting from the movement of people and goods have increased at a faster pace than for other sectors (see Figure 1.1). By 1997, transport emissions had increased as a share of emissions in almost all countries, compared with 1990 as well as with 1980 and 1971.

Figure 1.2 shows how per capita carbon emissions from passenger and freight rose along with per capita GDP from 1971 to 1997 for seven IEA

Figure 1.1. World CO_2 Emissions from Fuel Combustion

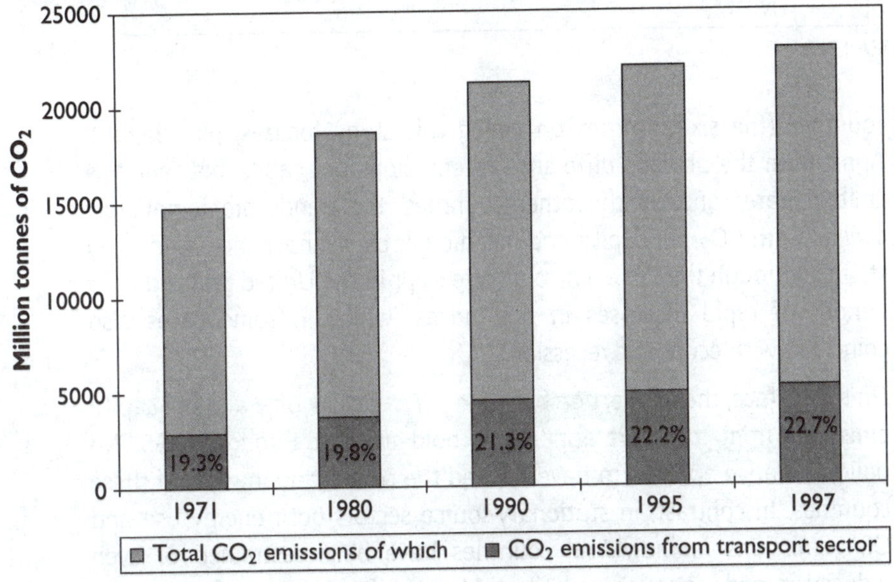

Source: IEA

Figure 1.2. **Per capita GDP and per capita Carbon Emissions from Passenger and Freight Transport, 1970-1997**

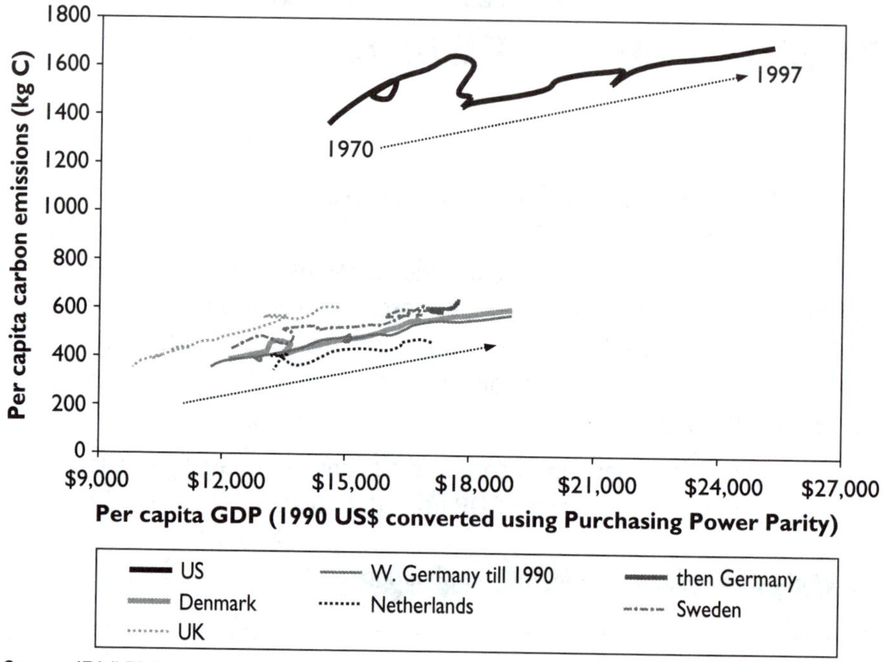

Source: IEA/LBNL

countries (the six countries on which this study focuses, plus Japan). Apart from the obvious difference in emissions per capita between the United States and all the other countries, the trends are remarkably similar. Both CO_2 per capita and income per capita have increased fairly steadily through time in each country, except in the United States during periods of rapid increases in fuel prices (which in some cases also coincided with economic recession).

On the surface, the connection between per capita income and per capita emissions from transport appears to hold at least through 1995. No policy initiative appears to have severed the connection in most of these countries. In contrast, in stationary-source sectors both energy use and CO_2 emissions in individual countries have been restrained or even reduced in end-uses such as boilers. Major shifts occurred after the two oil crises of the 1970s and the subsequent efforts to save fuel. The steady upward trend through these same periods in CO_2 emissions from transport suggests that restraining carbon emissions from transportation may be more difficult than in other sectors. Whether the efforts undertaken since the signing of the Kyoto Protocol can finally "sever the connection" in transport remains an open question.

Where to Look for CO_2 Savings?

The Kyoto agreement brought focused attention on a set of straightforward, but difficult-to-answer questions. How much can CO_2 emissions be reduced in each sector in the near term (2008 to 2012)? How much would such reductions cost? Which policies could gain enough political or popular support to be implemented?

Policy-makers in developed countries are now looking for real-world answers to these questions. Policy-makers in many countries have begun to question even the possibility of *any* CO_2 reduction, particularly in the transport sector. They point to the apparent high cost and low popularity of many potential transport policies and measures. This scepticism exists despite studies that suggest an abundance of cost-effective CO_2 reductions through a variety of measures.

Why will it be so difficult to restrain or reduce emissions from the transport sector? For one thing, strong and complex forces drive increases in transport activity, and hence CO_2 emissions. Within the transport community, there is wide uncertainty over such elementary parameters as the price and income elasticities of car ownership, levels of transport activity, and fuel consumption. Further, transport is closely linked to all other sectors of the economy. Many factors that influence transport lie outside the control of transport, energy, or environmental ministries. Moreover, many transport policies are set at state and local levels, not national. Others can *only* be implemented at the supranational level, but this requires hard-to-achieve international co-operation.

CO_2 Reductions in the Context of Transport

Emissions of carbon dioxide are not the only important problem transport raises. Others include safety, pollution, competition for urban space, and energy-security risks associated with increased reliance on imported oil. While transport systems provide enormous benefits to every economy, there are many situations where such "external" costs are also high. Frequently, they are of greater concern to policy-makers than the destabilisation caused by CO_2 emissions. Looked at more positively, reductions in CO_2 emissions can be realised as a fringe benefit of actions taken to reduce some of these other societal costs. This was emphasised in a recent study organised by the European Conference of Ministers of Transport (ECMT 1998), which concluded that the "internalisation" of such costs, through taxes and or other fixed measures, could have a significant restraining impact on CO_2 emissions. The ECMT's report of the estimated percentage of GDP lost to different external effects of transport is shown in Figure 1.3.

Figure 1.3. Average Estimates of the Total External Costs of Road and Rail Transport

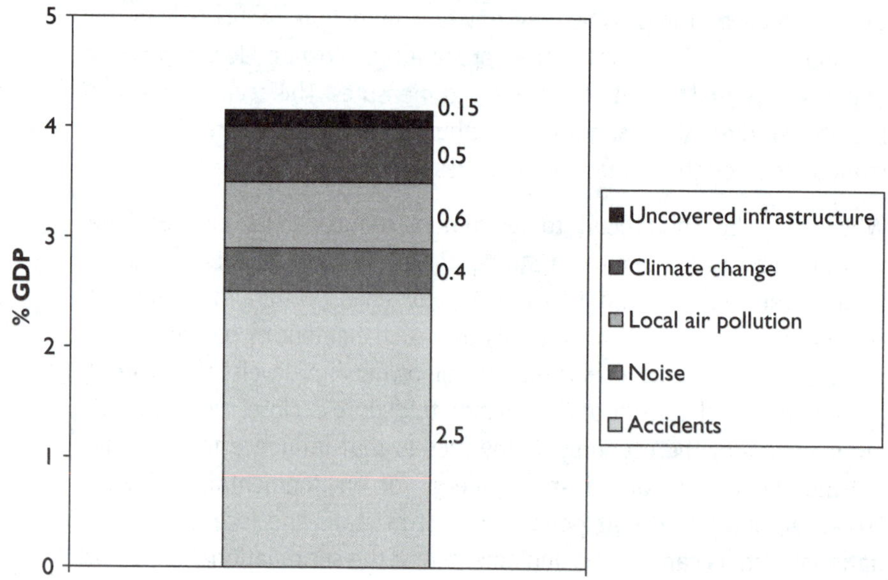

Source: ECMT 1998

Because CO_2 is affected by the amount of travel, the average fuel intensity of travel, and the choice of fuels used, a wide variety of other policies, including nearly all transport policies, affects CO_2 emissions. But to obtain large-scale reductions in CO_2 emissions in the future, policies may need to be adopted to target CO_2 emissions more directly. The extent to which national policy-makers have begun to do that is a central concern of this book.

Unlike other external effects of transport systems, climate change may not appear to be a compelling problem for the present generation. And there is still considerable debate about the timing and extent of the damage that will result from current greenhouses gas emissions. So, there may be limited political will to undertake painful measures to restrain CO_2 emissions. Still, policy-makers in most countries are under pressure from certain constituencies to slow or reverse transport's rising CO_2 emissions *now*. Countries also differ in the number and strength of those who *oppose* strong action to reduce CO_2 emissions. This book considers the roles of such groups, and how the interaction of various parties has influenced policy-making in each of the studied countries.

CHAPTER 2: ANALYSING POLICIES TO REDUCE CARBON EMISSIONS

This section presents a general framework for considering and categorising policies that affect CO_2 emissions from transport. Policies in each category are considered country by country in the following chapters. The framework should prove useful in considering where each country is placing its emphasis and how individual policies can be (and often are) combined into effective packages. While the framework is useful in developing a detailed quantitative analysis of the impacts of different measures, we do not attempt such an analysis in this volume. However, other upcoming IEA publications (including one on fuel economy improvement being published concurrently with this report) will provide such quantitative estimates for a wide range of transport policies and measures. Other recent studies have also done this – from the governments of Sweden (SIKA 1999), Denmark (ENS 2000), and the Netherlands (RIVM and VROM 1999) – and an earlier study from the US Energy Information Administration (Kyoto Scenarios 1997).

Linking Transport with the Environment and CO_2: the Decomposition Approach

The links between transport, energy use, and CO_2 emissions can be characterised using the simplified analytical framework illustrated in Figure 2.1. In this approach, CO_2 emissions are equal to the product of transport activity *A* (measured as passenger-kilometres or tonne-kilometres), modal structure *S* (the share of each activity by transport mode), modal energy intensity *I* (energy use per unit of passenger or freight travel by mode), and the emission rate *F* (CO_2 emissions per unit of energy consumed). This is called the "ASIF" decomposition.

Figure 2.1. Components of Transport that Determine Carbon Emissions

$$G = A * S_i * I_i * F_{i,j}$$

G	A	S_i	I_i	$F_{i,j}$
Carbon Emissions from Transport	Total Activity	Modal Structure	Modal Energy Intensity	Carbon Content of Fuels

i = transport mode, j = fuel type

Based on this methodology, there are four ways for governments to intervene, through policies, new technologies or both:

- Reduce or restrain the growth of activity *(A)*, the movement of people and goods.
- Shift traffic to less energy-intensive or carbon-intensive transport modes *(S)*.
- Reduce the modal energy intensity *(I)* of the various modes – using less energy for the same activity or getting more activity from the same amount of energy; this can be achieved by improving the vehicle technology (using less fuel per kilometre), improving utilisation (carrying more passengers or tonnes of freight per vehicle-kilometre), or improving traffic conditions so that vehicles perform better.
- Reduce the CO_2 content of fuels themselves *(F)*.

In the past, increases in emissions have stemmed mainly from increases in activity *(A)*. Growth in transport activity has been tied to income growth, more or less tightly depending on the country. This is illustrated in Figure 2.2, which shows the relationship between per capita GDP and the amount of passenger car travel per capita in a selection of IEA countries.

Shifts to modal structure *(S)* of relatively high energy intensity, such as passenger cars, have also contributed to increases in CO_2 emissions. So have reductions in load factors *(CU)* and trends in vehicle characteristics *(VC)* that have increased (or have slowed reductions in) modal energy intensities *(I)*.

Care should be taken in using this framework to categorise policies. For one thing, the four components that determine total transport emissions

Figure 2.2. Vehicle-kilometres Travelled per year per capita and GDP per capita in a Selection of IEA Countries, 1970-1997

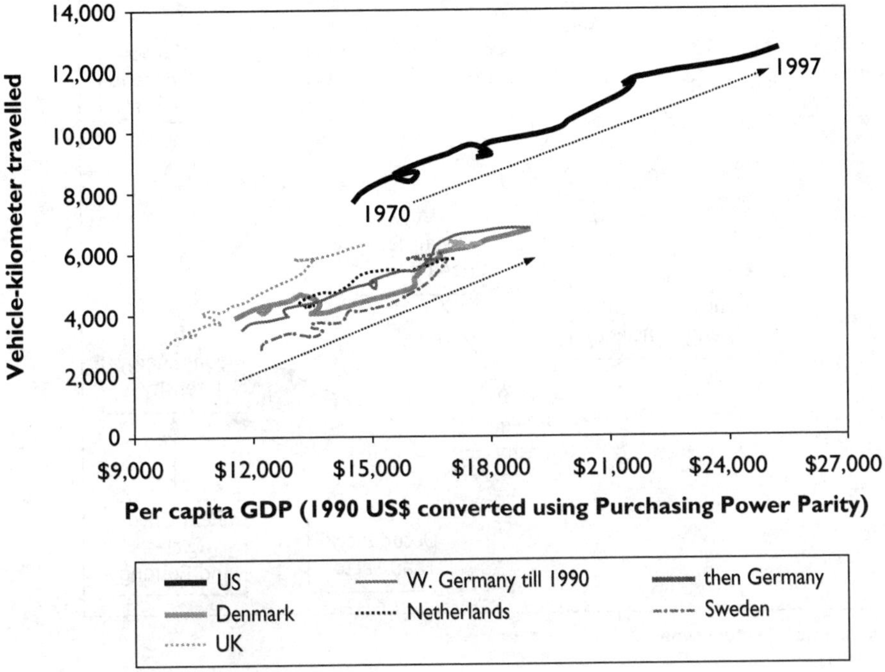

Source: IEA/LBNL

are not necessarily independent of one another. A change in one term may trigger changes in other terms that offset part of the effect of the first change. For example, reductions in modal intensity may reduce the cost of travel and thus lead to increases in transport activity. This is known as the "rebound effect."

It cannot be assumed that similar CO_2 reductions can be achieved for each component, or at the same cost, or over the same time frame. Each component may respond quite differently to changing prices for fuels or for vehicle use. Finally, the various policies that are available to obtain reductions in each of these components may have quite different political acceptability.

Figure 2.3. Actions in IEA Countries to Restrain CO_2 Emissions from Transport

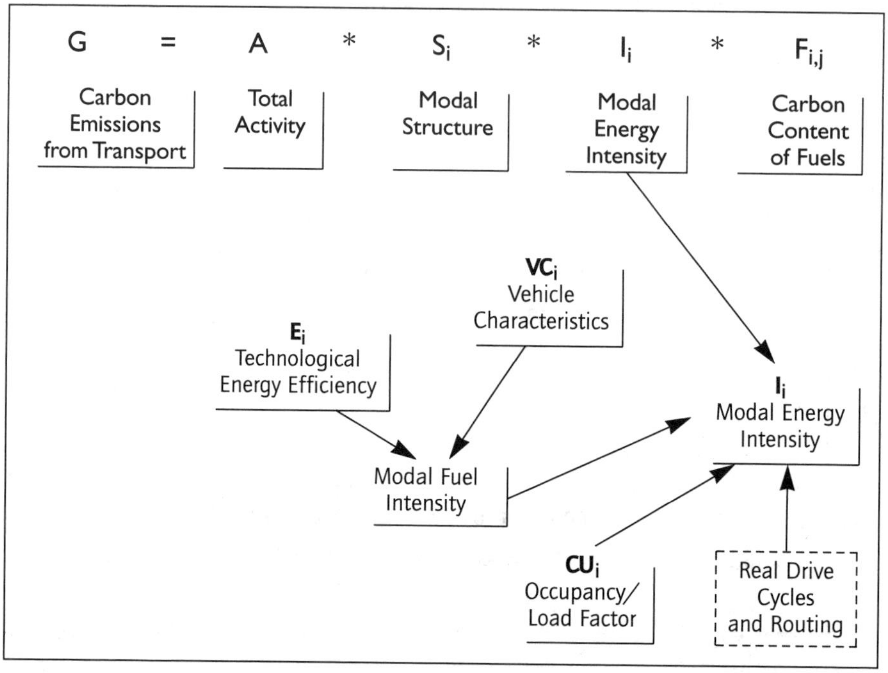

i = transport mode, j = fuel type

The time needed for policies to have their full effect is often longest for those policies that have the biggest impact. In the short term, activity and modal mix can be most easily modified, although it may take strong policy pressure to really change these components relative to their underlying trends. Further, the changes may not be permanent or may decline over the long run unless policy pressure is maintained or increased.

Technological advances cannot usually be implemented on a large scale in the short term because it is far more cost-effective to make these changes only to new vehicles. New features enter the fleet only over time, as vehicle stock turns over. In the longer term, however, technology improvements can dramatically reduce CO_2 emissions per vehicle-kilometre travelled; the introduction of hydrogen fuel cells is one example of this potential.

Although the potential for CO_2 reductions from change in fuel intensity is large, two key factors influence the rate of possible change: the rate of replacement or growth in the vehicle stock, and the difference between the average fuel consumption of existing cars and that of new vehicles. If vehicles are being replaced rapidly or the stock is growing rapidly, then average fuel intensity can change fairly quickly. In the 1990s, however, stock replacement in many countries slowed down. This was a natural consequence of slowing growth in the number of people of driving age and near-saturation in vehicle ownership rates. Although national policies have led car manufacturers to try to reduce new-car fuel intensity (Figure 2.4), the results in term of on-road car fuel intensity in each

Figure 2.4. New-car Test-fuel Economy, All Fuels (sales-weighted test values) in Selected IEA Countries, 1970-1999

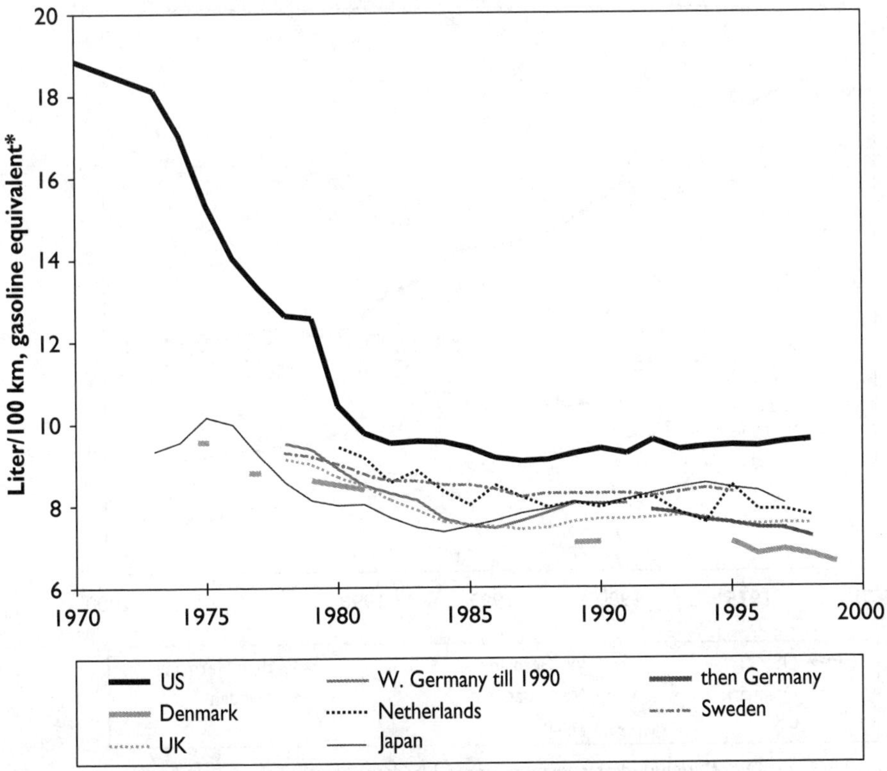

* Gasoline, LPG and Diesel included at energy content. US data include light trucks/vans
Source: IEA/LNBL

country is modest or negligible (Figure 2.5). Improvements in vehicle quality and increases in the number of vehicles owned by individual families have also stretched the real lifetimes of vehicles.

Technological progress can lower fuel consumption rapidly in new vehicles, but there should be sufficient demand for advanced-technology, high-efficiency vehicles to justify the investments needed to produce them. Because such advanced cars often cost much more than others, generating demand can be difficult. And to provide actual reductions in fuel use and CO_2 emissions per kilometre, efficiency improvements should be made at a faster rate than increases in vehicle size, weight and

Figure 2.5. Road Car Fuel Intensity in Selected IEA Countries, 1970-1998

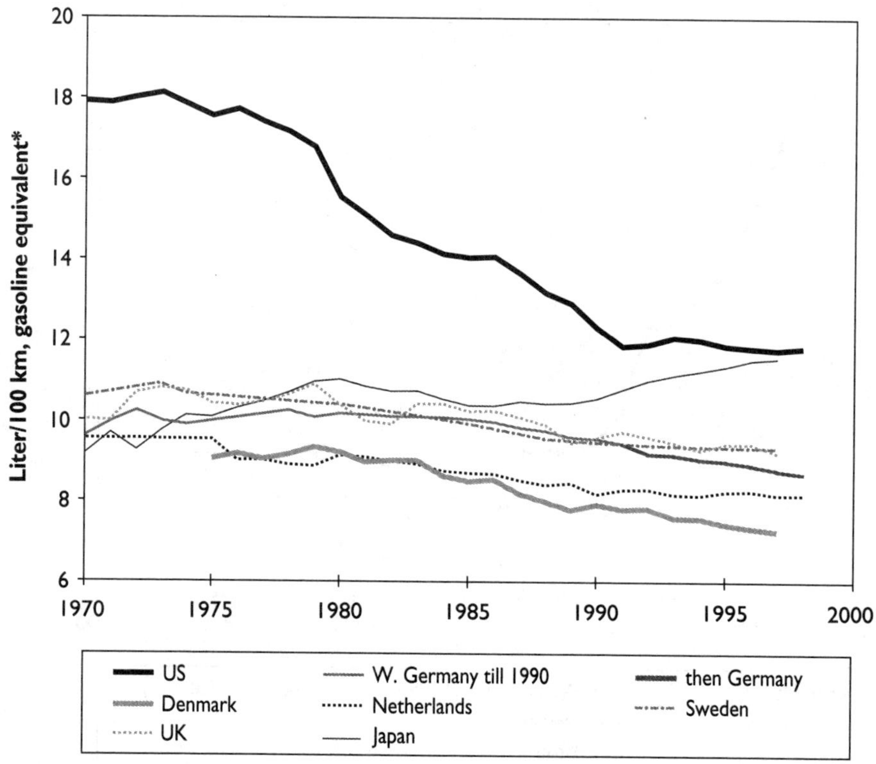

* Gasoline, LPG and Diesel included at energy content. US data include light trucks/vans
Source: IEA/LBNL

power. Over the past ten years, these factors have tended to offset the fuel-reduction benefits of incremental technology improvements.

Policy Categorisation in the Decomposition Framework

Once the main components affecting CO_2 emissions are identified, one should then identify the policies and measures available to change these components. For each component of Figure 2.3, one should ask what kind of policy pressure might be applied, over how much time, how strongly, and with what result.

Transport policies can be grouped into eight general categories representing different "points of intervention." Policies applied at these points affect different components of the ASIF diagram, with some affecting more than one component. The points of intervention are the following:

- **The costs of fuel,** which can be increased by various kinds of ad valorem or quantity taxation; taxes can be adjusted to reflect likely pollution from using these fuels, or simply the energy or carbon content of the fuel.
- **Other variable costs of motor vehicle use,** where charges could be levied on total distance driven, on using cars at peak times or in congested areas (road pricing or zone pricing); insurance premiums could be set that are directly proportional to vehicle use; charges could be imposed on trucks on the basis of their payload.
- **The conditions of road traffic flow,** which can be improved by electronics and signalling technology, or by restricting or adding capacity. The latter has been the principal element of transport policy in most countries since World War II.
- **Public transport and other alternatives to road transport,** which can be made more attractive by reinforcing schedules, building more nodes for transfers, improving the comfort of vehicles, lowering fares, or offering incentive fares.

- **Vehicle production (supply),** which can be improved mainly through research on vehicles, their propulsion systems and on new fuels themselves.

- **Vehicle fleet demand and characteristics,** which can be moulded through changes in vehicle taxation or regulations on vehicle fuel, weight, power, fuel consumption, emissions of pollutants and so forth; by labelling of the fuel economy of new vehicles; by changes in the conditions of vehicle use that lower the marginal benefits of using cars (street closings, traffic calming); or by developing schemes like car-sharing that make *owning* cars less important.

- **Urban space,** where relevant policies include regulations on where new homes or other buildings can be put, regulations on provision of parking, restrictions on traffic in certain neighbourhoods, or changes in property taxation.

- **Public attitudes towards transportation,** which can be influenced by education and information campains.

It will be useful to now examine which policies have the most impact on which components. This is what we seek to do in Table 2.1, which is intended to serve as a point of reference in the discussions of actual policies in different countries. One point worth noting is that there are quite a few null sets – cases where a given policy has no impact at all on one or more components. Thus, the particular area of transport that is being targeted will probably determine which type of policy it makes sense to consider.

Other Important Aspects of Policy Analysis

In the policies and packages of policies reviewed in the following chapters, a number of aspects differentiate one country's situation from another. These differences may affect the paths chosen by different countries.

The Past. What is the history of CO_2 emissions reduction in each country? How many programmes had already been implemented by 1990 that are now imbedded in the baseline? What additional measures appear to be available to countries that have been tackling the problem over several decades? Recent trends in underlying components are also important. How fast is travel growing? Are there signs of saturation in vehicle ownership or use that might help countries in their efforts to sever the connection between growing wealth and growing carbon emissions?

Public and Institutional Acceptance of Policies. What types of institutional and democratic constraints should each country's government deal with when developing transport/CO_2 policy? We identify the obvious, and in some cases the subtle, political difficulties or advantages, such as the presence or absence of a vehicle industry, the strength of lobbies, the tradition of government intervention, and basic public trust (or mistrust) of government. In fact, the governments of *all* countries covered in this study operate under very tough constraints, including public resistance to increased marginal cost of travel, resistance to measures that hurt specific industries and the general need to address a broad variety of social and political objectives that overlap and in many cases conflict. In the case of CO_2 emissions, most policies have a strong impact on something else, such as transport cost. Thus, the policies most likely to be successful are those that provide multiple benefits – reductions in other emissions, improvements in safety or increases in jobs, as well as carbon reductions.

Underlying Economic Conditions: Incomes, Fuel Prices, Fuel and Car Taxation. Initial conditions can determine the political feasibility and the actual effects of new CO_2 reduction policies. For example, fuel prices are relatively low in the United States, which means that even a modest increase in fuel taxes (such as 5 cents per litre) would be very visible there, while the same tax hike would hardly be noticed in Europe (at least before the retail price protests of September-October 2000). Conversely, fuel taxation is so high in Europe that authorities in many countries establish significant differentials among fuels (between leaded and unleaded gasoline, or among gasoline, diesel, and LPG) in order to rapidly

Table 2.1. Policy Classification

	Policy group	**Policies to change price of fuel**		**Policies to change other variable costs**	**Policies to influence traffic flow**		**Policies to enhance public transit**
	Example	*Fuel Tax Carbon Tax*	*Others*	*Road pricing; parking charges; varying other costs*	*Highway building, Intelligent Transport Systems (ITS), computerised traffic management)*	*Restraining traffic flow (traffic calming, traffic diversion, speed limits)*	*Expansion of service to reduce travel time and/or wait time, expansion of service area, better comfort, reduction of transit fares or regulatory reform to encourage private sector investment*
	Effect on:						
A	Activity	Slight restraint, because low elasticity		Slight restraint short term, possibly more in longer term	Induces activity	Restrains activity	May induce activity
S	Structure	Slight shift, cross-elasticities low		Increasingly favourable to collective and non-motorised modes	Favours cars and trucks	Favours buses and bicycles	Favours collective modes for existing trips; could result in overall increase in car travel
I	Engine energy intensity	Possible reduction in medium to long-term		No impact	No impact	No impact	No impact
	Vehicle size	Possible reduction in medium to long-term		No short run impact, long-term impact depends on policy	No impact	No impact	Small impact; possible increase in bus/train size
	Capacity Utilisation	Possible increase (short and long-term)		Possible increase (short and long-term)	Possible decrease in utilisation	Possible increase in utilisation	Depends on policies adopted; oversupply may lead to reduction in utilisation
	Operation Optimum	Improvement only if important impact on **A** or **S**		Improvement only if important impact on **A** or **S**	Improvement	Deterioration	May improve if significant modal switching occurs
F	Fuel Mix	No impact	Favours low-carbon fuels	No impact	No impact	No impact	May improve overall fuel mix, depending on policy

Policies to influence vehicle fleet demand	Policies to influence vehicle fleet supply/production	Policies to influence urban structure	Policies to influence public attitudes towards transport and energy consumption
Fees on acquisition, ownership, and registration to influence ownership decisions; taxing company-car benefits as ordinary income; revenue-"neutral" cross-subsidy schemes such as "feebates"	*Advancing vehicle fuel economy or alternative fuel technology via: support for manufacturer research and development; sponsorship of independent R & D; regulations and performance mandates; tradable permits*	*Land-use controls for urban development; co-ordination of transportation and land-use development; tax benefits for choosing to locate in "accessible" parts of metropolitan regions; enterprise zones; location-efficient mortgages*	*Media campaigns, youth education campaigns, information "exchange" projects*
No impact	No impact	Little impact on trip rates; reduction in trip distances	Potential improvement
No impact, except for car-sharing schemes (favour buses and bicycles)	No impact	Favours collective and non-motorised modes	Potential improvement
Reduction in medium and long-term	Potentially important reduction with Turbo Direct Injection (TDI), Gasoline Direct Injection (GDI), Continuously Variable Transmission (CVT) technologies	No impact	Potential improvement
Potential improvement, particularly under car-sharing schemes	Some reductions available with materials, Air conditioning tech., etc.	No impact	Potential improvement
Potential improvement, particularly under car-sharing schemes	No impact	Possible improvement, if policies render collective transport or carpooling more attractive	Potential improvement
No impact	No impact	Unclear; possible deterioration	Potential improvement
If set up right, could improve markedly	Short-term potential with low-sulphur diesel or hybrids, medium-term with battery-electric, long-term with hydrogen fuel-cell electric technologies	No impact	Potential improvement

affect fuel choices for new cars. This may not be a viable approach in low-tax countries.

Focus on Nearer-Term or Longer-Term Policies. Does a given country rely on measures to reduce activity or to switch activity from one mode to another, measures which can provide fairly fast results? Or does it focus on technological measures, which may take much longer to have an impact but which may ultimately produce greater CO_2 reductions? Or does it appear to strike a balance between these approaches? Large market countries with a domestic auto industry are better positioned than others to implement policies that affect technology development and its adoption.

The real policy challenge is twofold. Countries, or international bodies, need to overcome heavy opposition to policies to reduce CO_2 emissions. And the required changes in technology and behaviour should be great enough to offset most or all of the natural growth in overall traffic. Indeed they must go well beyond that if they are to provide significant absolute reductions in CO_2 emissions.

CHAPTER 3: DENMARK

Background

Trends in Transport and CO_2 Emissions

Relative to many other countries in western Europe, Denmark is "under-motorised" for a country with its high GDP. But, the average car in Denmark is driven more than the average car in any other country in western Europe except Finland (roughly 33 percent more than in Sweden and 45 percent more than in Germany). So levels of vehicle travel per capita are about average for western Europe. This point taken with above-average levels of bus and rail use, makes the Danes among the most mobile citizens of Europe, despite their low car-ownership rate.

Due primarily to very high purchase and registration taxes, cars in Denmark are relatively small, even for Europe, and have correspondingly low fuel intensities.

Buses and trains account for nearly 30 percent of motorised travel, and the shares of bicycling and walking are also well above average, accounting for nearly 10 percent of total travel (a record similar to that of the Netherlands). The combination of low fuel intensities in cars and extensive use of buses and trains explains why Danish CO_2 emissions from travel are among the lowest in Europe, despite the above-average overall amount that Danes travel.

In freight transport, Denmark represents something of an anomaly. Freight volume has generally increased with GDP growth, but at a lower rate than in other countries. Truck fuel consumed per tonne-kilometre, however, is among the highest in the IEA. Denmark consumes a large amount of energy for freight relative to its GDP because freight is mainly shipped by trucks.

Table 3.1. Key Transport Statistics for Denmark, 1997

Population (million) GDP per capita (US$ in PPP)	5.2 26,280	Cars (per 1,000 people)	338
Passenger transport activity (billion passenger-kilometres)	**82.9**	**Freight transport activity** (billion tonne-kilometres)	**20.1**
Passenger car	*79%*	*Road*	*73%*
Powered two-wheeler	*1%*	*Rail*	*8%*
Bus	*14%*	*Inland navigation*	*0.0%*
Railway	*6%*	*Oil pipeline*	*19%*
Transport final energy consumption (million tonnes of oil-equivalent)		**Transport CO_2 emissions** (millions tonnes)	
1990	**3.8**	1990	**11.4**
1997	**4.0**	1997	**11.8**
Rail	*3%*	*Rail*	*2%*
Road	*93%*	*Road*	*93%*
Inland navigation	*4%*	*Inland navigation*	*4%*

Source: *EU Transport in figures*, OECD, IEA

Figure 3.1. Trends in CO_2 Emissions in Denmark, 1970-1997

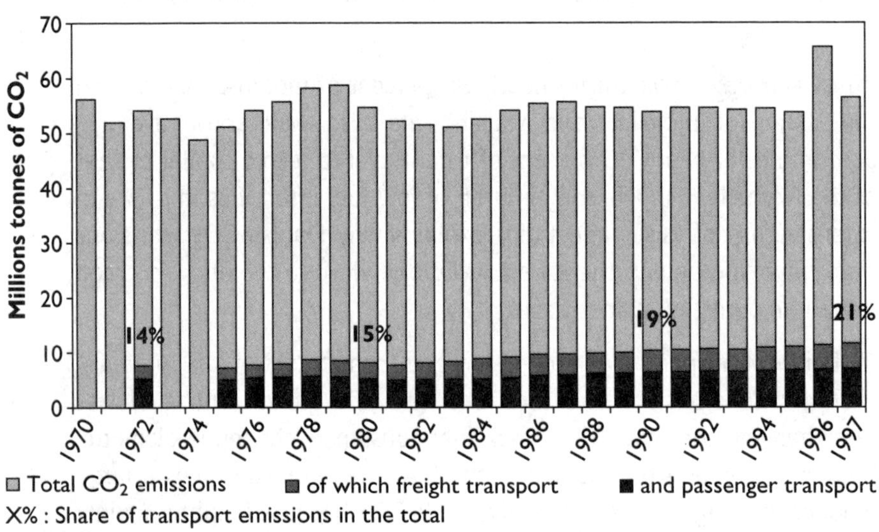

Source: IEA/LBNL

A recent study by the Danish Ministry of Transport indicated that by 1999 transport energy use and CO_2 emissions in Denmark were growing well beyond original projections, but that this growth is expected to slow – if not be reversed – by the middle of the present decade (Ministry of Transport 2000). Table 3.1 gives some key statistics for the Danish transport sector.

Figure 3.1. shows trends in Danish CO_2 emissions from 1970 to 1997.

Public Attitudes and Perceptions

Danes have accepted more than the usual level of government influence to discourage auto use. The present energy/environment minister recently celebrated 25 years in the parliament (Folketing), a testimonial to the public's approval of the country's policies over this period. In this atmosphere of support, the energy/environment and transport authorities have united behind the Ministry of Transport's CO_2 plan (Trafikministeriet 1997). But some disagreements remain mainly about the pace at which fuel prices should be raised. Consequently, only a limited number of the proposed policies has been put in place.

Roles and Position of Industry

Denmark has no domestic car industry. There are not even any "transplanted" factories in Denmark operated by foreign car companies, as there are in the Netherlands. The main domestic actor on the automobile industry side, the Association of Car Importers, has taken a public position in favour of lowering taxes on new cars. The association commissioned a study by the consultant company COWI that supports its position on new-car taxation (COWI 1995b). Recently, the government initiated a study of such reform (Trafikministeriet 2000). The government has invited the three main automobile importers – Danish Auto Importers, Automobile Agents of Denmark and Danish Motorist Society – to participate in discussions about developing a tax scheme based less on car value and more on environmental considerations. Although the lack of a strong auto industry voice in Denmark means that there may not be much industry resistance to reform, it also means that the Danish government, acting alone, cannot stimulate major changes in manufacturer behaviour, given the small market.

Other Influences and Trends

A key "outside" influence on policy in Denmark is the Ministry of Finance. Denmark raises a significant amount of overall revenue from taxes on transport, an amount estimated at more than US$1,000 per capita in the early 1990s (Schipper and Eriksson 1995). The 1999 finance bill introduced by the Ministry of Finance stresses the connections between travel, emissions, and taxation. It points out that the increases in gasoline prices in 1993 (motivated by a large-scale tax reform that included environmental taxation) produced a slight increase in the share of collective transport in Denmark, just as a decline in fuel prices after 1982 caused a fall in collective transport's share.

Evolution of Transport-CO_2 Policy

In several respects, the transport policy situation in Denmark is atypical for countries in western Europe. Since World War II, purchase and registration taxes on cars in Denmark have been among the highest anywhere. The registration tax on new purchases is 180 percent of vehicle price (and is more than 200 percent if the 25 percent VAT is included).

After the oil shock of 1973, Denmark took a number of actions to reduce oil dependence. The Danes implemented a large-scale retrofit programme to reduce fuel consumption for space-heating needs (where oil was by far the dominant fuel source). Also during the 1970s and early 1980s, Denmark converted its power sector from near total dependence on oil, substituting coal as the major fuel. Denmark also raised taxes on many forms of energy during this period.

As a result of these changes, coupled with development of its North Sea oil reserves, the country became nearly self-sufficient in oil and gas. The

Danish government still taxes energy use by individual households more than any other government in Europe, and it taxes the purchase, ownership and use of transport vehicles at a higher level per capita than any other country except Italy (Schipper and Eriksson 1995; Trafikministeriet 2000). Danish gasoline taxes were lowered somewhat during the 1980s, in part to discourage Danes from buying gasoline in nearby Germany and to comply with EU directives. But these taxes have crept back up in recent years.

In addition to bringing oil production in line with consumption, Denmark's policies have resulted in very low car ownership, especially in view of its citizens' high incomes. Yet these light taxes have not measurably restricted travel per capita in the country, which is among the highest in Europe. Even passenger car travel is above average, since the Danes compensate for lower car ownership by driving each car farther.

Perhaps the bigger impact of the vehicle-tax policy, at least in terms of carbon emissions, has been on the size and fuel economy of vehicles, which is better than average – but not the best – in western Europe. High taxation *has* played a key role in encouraging transit use, biking, and walking. Other policies by the Danish government have, of course, also played a role in promoting these environment-friendly modes.

In freight, Danish energy use and CO_2 emissions per unit of GDP are among the highest in western Europe. As in most EU countries, diesel fuel has not been taxed commensurately with gasoline. In Denmark, professional truckers receive a refund of 100 percent of the diesel tax they pay. This has led to some misuse of road diesel fuel for space heating.

In 1990, Denmark began a series of analyses and planning exercises that have carried through the decade, resulting in numerous reports and plans. An ambitious plan was developed by the Ministry of Transport in 1990, as described in the *Transport Action Plan for Environment and Development* (Trafikministeriet 1990), and *Energy 2000,* the National Energy and Environment plan, was produced by the Ministry of Energy (Energiministeriet 1990). In this process, energy and environmental authorities began working more closely with transport authorities to develop plans to fight air pollution and traffic problems, as well as CO_2

emissions. Denmark's concern for CO_2, even though not the primary driver in its plans, pre-dates the "Earth Summit" (United Nations Conference on the Environment and Development), held in Rio de Janeiro, Brazil, in 1992. In 1993, the Ministry of Transport put forward a new transport plan (Trafikministeriet 1993). In 1996 came a Government Parliamentary Proposal, which was finalised as the Ministry of Transport's March 1997 *CO_2 Reductions in Transport* (Trafikministeriet 1997). The most recent summary of transport trends and policies is to be found in the Ministry of Transport's *Transport Policy Review* of February 1998.

These reports and others (for example, COWI 1996 and Trafikministeriet 1995a and 1995b) have put strong emphasis on quantifying and addressing "externalities" in transport. This emphasis is especially evident in the series of reports produced by the consultant company COWI, which offered a comprehensive and systematic approach to CO_2 reduction across transport sectors and policy types. This is important, because the key assumption behind all the Danish efforts is that the benefits from improving transportation and from reducing CO_2 are greater when considered together than if the benefits of CO_2 reduction were considered alone. One key study shows that the value of resolving other problems in the transport sector is so great that the CO_2 reductions, which may be appreciable, became almost "free" (COWI 1994, 1996).

The result of a decade of analysis and planning initiatives is a set of policies, largely oriented towards "getting the prices right." But many more policies have been discussed, and are still being considered, though they have not yet found their way into law.

While a new global initiative has not been formally adopted by the national government, and many of the proposed initiatives are still under discussion, some of the elements of various packages, as well as of the transport-CO_2 proposal of 1990, have already been put into place. These include the following:

- The main loopholes in diesel fuel pricing and truck taxation have been addressed (Transportraadet 1995).
- New purchase taxes on light trucks and vans have been shifted to reflect weight or capacity, to encourage the purchase of smaller trucks.

- Lower prices for mass transit and new taxes on air travel were announced and put in place in May 1997.
- The "Green Owner" annual registration fee scheme came into effect at the beginning of 1998 on all cars bought in the spring of 1997 or later (see section below), shifting some of the previous weight-based fee to a fuel consumption-based fee. The plan was expanded to cover diesel cars in 1999.
- Changes in the system of new-car taxation, which had been entirely *ad valorem*, were enacted in 1999 and provided a considerable tax break for new cars with fuel intensity lower than 4 litres per 100 kilometres. However, only one vehicle model available in 2000 meets this criterion (Faerdselstyrlsen 2000).
- Enactment of a new-car fuel-economy labelling law, effective from 1 March 2000 (Faerdselstyrlsen 2000).
- Fuel taxes were recently increased as part of a fiscal package designed to slow consumer spending. The increases were DKK 0.35 per litre (€ 0.046 per litre) from 1994 to 1997, and DKK 0.50 per litre (€ 0.067) on 1 January 1999, to be followed by DKK 0.125 on 1 January 2000 and 1 January 2001, as well. For diesel, the tax increases on 1 January 2000 were DKK 0.235 per litre, and the difference between taxes on low-sulphur and ordinary diesel was maintained at DKK 0.18 per litre.
- A small yearly CO_2 tax increase has been put into place. (By 1997, the real price of fuel was back to where it was in the early 1990s.)
- A mandatory inspection of all cars over four years old was introduced for safety reasons. This may indirectly act to remove the dirtiest and most energy-intensive older vehicles from the road.

A number of proposals have been, and are still being, considered but have not, as yet, been enacted:

- Strengthen land-use planning by limiting the growth of built-up areas, restricting out-of-town shopping centres, strengthening bus and cycle lanes.

- Improve driving conditions through computerised traffic direction and parking-place instructions, telecommuting, speed limits enforced by more highway police and improved driver education.
- Emphasise improved trucking practices, reduced volume of packaging, some modal shifts or combined transport, and increased registration fees on medium-size trucks for local delivery (in order to promote use of smaller trucks).
- Encourage development of biofuels, principally biodiesel and ethanol, or shifts to natural gas or DME (a fuel made from natural gas and biomass), although this is now regarded as less important than encouraging electric vehicles.
- Improve the quality and speed of local collective transport, coupled with lowering fares on those where direct competition with automobiles exists.
- Encourage shifting car trips of under 3 kilometres to bicycles, which already (along with walking) account for about 6 percent of total travel in passenger-kilometres.

New proposals were made in 1995-97 by the Ministry of Transport, based principally on those given above, but expanded through a series of studies. The goal of this package was to reduce CO_2 emissions by 2005 to their 1988 levels, focusing for the most part on passenger cars. Key elements of the proposals included:

- A voluntary agreement with the Car Importers Association to aim for 5 litres per 100 kilometres of gasoline (4.5 litres of diesel) by 2005, differentiated fees on new cars, a rising fuel tax, and increased information on new-car fuel use. This measure has become very important in the eyes of the Danish authorities, but the issue is very much out of their hands for now, as a EU-wide agreement has set a goal of an equivalent of 6.2 litres per 100 kilometres of gasoline.
- Alternatively, a requirement to reduce new-car fuel intensity by 20 percent from 1998 averages by 2005.
- A tax to keep the cost per kilometre of driving at least constant as the fuel intensity of new cars falls.

- A slightly lower first-time registration fee for new cars (currently an *ad valorem* tax at 180 percent of the pre-tax price for the portion of price above about US$6,000, 105 percent for the portion below US$6,000) to compensate for the slightly higher costs of more efficient new cars.
- A shift in yearly registration taxes away from car weight and towards initial test-fuel consumption (the "Green Owner" fee mentioned in section III), with electric vehicles being exempt.
- Shifting to a new-car tax based more on fuel consumption than on vehicle price. One proposal would increase the tax on any car with 10 percent higher than average fuel intensity, rising to a DKK 10,000 (€ 1,339) penalty on cars that use 20 percent more fuel than the average. This would complement the vehicle-tax reduction for cars with very low fuel intensity implemented during 1999.
- New-car labels that indicate test-fuel consumption and CO_2 emissions.
- A re-evaluation of the way collective transport modes are priced, to be undertaken in 2000.
- Per-kilometre fees (road pricing) for trucking (not part of the 1995-97 plan).
- Taxation of fuel used by aircraft and ferries (not part of the 1995-97 plan).

As of early summer 2000, few of the newer elements of this package had been passed by the Danish Parliament, but the basic bill, the Government's Strategy for Reduction of CO_2 Emissions from Transportation, had been accepted by Parliament. Labelling took effect on 1 March 2000, and the first new-car "feebate" in Europe was enacted, lowering taxes on very fuel-efficient cars, which at first included only the Volkswagen Lupo diesel.

Selected Transport-CO_2 Initiatives

Green Owner Fee

The "Green Owner" fee policy came into effect in the beginning of 1998 and has aroused interest around Europe. This annual registration fee, which is separate from the very high one-time vehicle registration fee for new cars, replaces the previous weight-based system. It is based on both weight and the tested fuel economy of each model, and it uses a standard weight-based tax rate, plus or minus a differential based on rated fuel consumption. The 1998 differentials – that is, the amounts (more or less) which the owner pays each year relative to the previous taxation scheme – are shown in Table 3.2.

While the Green Owner fee represents an important step toward the use of fuel-consumption based fees, it has much less impact on the cost of vehicles than does the one-time new vehicle tax. Just consider the price of a new car including taxes. On average, the heavier and more fuel-intensive a car, the higher the new car price and the higher the new car taxes (up to 180 percent of vehicle price). The Ministry of Transport estimates that, for every litre per 100 kilometres a car uses, its taxed price increases by about DKK 45,000 (€ 6,025) (Trafikministeriet 2000). If the car travels 15,000 kilometres per year, it would require 150 more litres per year than a 1 litre/100km less energy intensive car. Over 10 years, the difference would be 1,500 litres. The extra cost of the new car, DKK 45,000, can be compared to the extra lifetime consumption of fuel, 1500 litres. In a sense, the extra fuel used "costs" about DKK 30 (€ 4) more because of the increased overall taxation of the car. However, all of this

Table 3.2. Differentials in Yearly Fee for Gasoline Cars in Denmark (January 1998)

If test-fuel intensity (litres per 100 km) is lower than:	5.55	6.25	7.14	8.33	10
Then the typical fee differential (DKK, converted to €) compared to previous weight-based fee is:	- 765 (€ 102)	- 430 (€ 57)	0	575 (€ 77)	1,375 (€ 184)

Source: Danish Energy Agency; Ministry of Finance.

Notes: The fee is part of the yearly registration cost and applies to cars bought after 1 January 1998. The actual fee is based on a continuous function using these fixed points. The balance point where the new fee equals the old fee will gradually move towards higher fuel economy. The values for diesel cars are more stringent, to reflect the greater energy and CO_2 content of diesel fuel and the larger number of diesel cars with lower consumption.

extra cost is paid at the time of purchase so it does not affect the level of travel after the car is bought.

Now let us consider the impact of the Green Owner annual registration fee. For the average driver, this yearly fee can add about DKK 3 per litre (€ 0.4) of extra fuel used, thus giving a small additional reminder each year of the energy intensiveness of the vehicle. One additional fee is still based on vehicle weight, which also sends a fuel-consumption signal to owners. Overall the new fee scheme is an important start toward shifting to a vehicle tax system based on fuel consumption (or CO_2).

In sharp contrast to the US or German approaches for gaining improvements in fleet fuel economy, the Danes clearly focus on the demand side of the vehicle-technology equation. To some degree, this strategy arises from necessity; since Denmark has no automobile industry of its own, supply-side policies would not work. But there is also a logic to the strategy that goes beyond simply responding to national conditions. In many car-producing countries, particularly the US, car manufacturers are reluctant to radically improve fuel economy rapidly, for fear that attendant changes in new-car performance (less speed and more power) will cause them to lose market share. This phenomenon logically suggests that co-ordinated demand-side measures are also needed to carry on this fight against CO_2 emissions. Denmark may provide a laboratory for the world to demonstrate how effective such demand-side measures can be.

Fuel Price Rises Greater than the Rate of Inflation

The Danes are strongly behind the EU agreement with car manufacturers aimed at cutting new-car CO_2 emission to 120 grams per kilometre by 2008. This would amount to a 20 percent improvement over current new-car averages for Denmark. The Danish proposal from 1995-97 includes regular increases in taxes to keep the cost per kilometre on a par with inflation – that is, to keep the real costs of driving constant. The purpose is to discourage people from driving more as vehicles become more efficient.

The way the proposal is phrased is a particularly important part of the Danish strategy. It does not specify or limit the nature of the price increases; it mentions simply an increase in the per-kilometre cost of driving. Increases may take the form of fuel price rises (taxes), road charges, or a combination of the two. In fact, because it is difficult to know exactly how consumers and manufacturers would respond to a fuel-price increase on top of the EU-sponsored voluntary agreements (cars may by then become even more fuel efficient), road charges are in many ways a more effective means of maintaining car-use parity.

The proposal as it stands may limit the price/rebound effect and dampen any rebound caused by better fuel economy. But, unless it is strengthened to raise the per-kilometre cost of driving by more than inflation, as opposed to simply keeping parity, it is unlikely to eliminate an income effect entirely.

Repeal of the Diesel Fuel Refund for Truckers

Between 1992 and 1994, Denmark gradually eliminated refunds of diesel fuel taxes to professional truckers. The system had amounted to a *de facto* subsidy to Danish truckers to the detriment of their non-Danish competitors, so it probably violated EU law. Its repeal, therefore, may not have been directly related to reduction of CO_2 emissions, but it has had a positive effect. The net effect was a rise of more than 40 percent in the price of diesel fuel within two years. Fuel use per tonne-kilometre of trucking has been falling.

In related measures, the purchase taxes on trucks will be revised to reflect their sizes. This will discourage the purchase of trucks larger than necessary, which, under the previous conditions of low real diesel prices, compounded the problems of truck fuel use that led to high fuel intensities. Empty running and low capacity utilisation were a large component of this problem.

Revision of Taxation of Trucks

Starting in 1997, the tax structure was changed. Under schemes in effect through 1997, medium-size trucks (2 to 4 tonnes) were taxed less for first-time registration than were light trucks and vans (less than 2 tonnes). Trucks over 4 tonnes paid no registration fees at all. After 1997, the fees for large trucks were raised, but the number of large trucks acquired changed little, while the number of new light trucks actually fell. Given the relatively poor capacity utilisation of trucks in Denmark, which has led to very high emissions per tonne-kilometre even after the rise in diesel prices in the early 1990s, the government is looking for a new approach.

Conclusions

Denmark's systemic approach to CO_2 and transport policy has produced a package of measures that do not isolate CO_2 emissions reduction as the only goal. CO_2 reduction is a natural result of a general policy to minimise the external costs of the transport system and rely on non-automobile travel as much as possible.

Because it relies so heavily on fiscal measures, transport policy is now as much the domain of the Finance Ministry as of the Environment and Natural Resources Ministry. The success of this multi-faceted CO_2 policy, and its strong focus on demand-side incentives, has been accompanied by good inter-ministerial co-ordination. In addition, where policy support at the EU level is required – particularly agreements with car manufacturers and fuel producers – common fuel taxation policy has proven to be effective and influential relations with Brussels important.

Another lesson to be drawn from the Danish experience is the importance of the balance between fixed and variable costs in the transport/environment equation. High costs for car acquisition and registration may indeed retard the growth of car ownership. But if variable costs are not also very high (and, perhaps, even if they are) households with cars will use them more frequently than in countries that have lower fixed costs and more vehicles per capita. In Denmark, high use per car has nearly offset the benefits of low ownership rates, resulting in roughly the same amount of overall car use as in other countries.

From the modelling work of COWI, it appeared that the ambitious goals of new government policy could largely be met. At present, however, only some of these goals have been moulded into policy. Moreover, as a small country with no domestic car industry, much of Denmark's success depends on what goes on beyond its borders. For example, if the goal of the European Union-ACEA voluntary agreement to reduce fuel consumption in new cars is met, it might provide significant reductions in the fuel intensity of cars sold in Denmark – but then again it might not. This is because the agreement is defined in terms of average fuel consumption across the EU. Since Denmark's fuel consumption is already well below average, it may not change as much as in some countries at the other end of the spectrum, such as Germany.

Denmark demonstrates what can be done in a relatively small country with no car producers, where the bulk of the population is prepared to accept the use of taxes and fees for the common good: high car and fuel taxation, and graduated taxes to encourage cars that consume less fuel. Nevertheless, of the numerous measures, some very bold, that have been taken since 1990, only a minority have survived the policy process. The most important are diesel price reform, trucking registration-fee reform and the Green Owner fee. Thus, the government has taken some steps, but a comprehensive package is still waiting in the wings.

CHAPTER 4: GERMANY

Background

Trends in Transport and CO_2 Emissions

Germany has among the highest car ownership and car use in Europe, though relative to income it is similar to other European countries. Compared with other European countries, car taxation in Germany has been modest, consisting only of VAT, and fuel taxation was sharply increased during the 1980s.

The fuel intensity of new cars decreased 20 to 25 percent between 1978 and 1985. The share of diesel vehicles in new-car sales increased from roughly 5 percent in 1978 to 18 percent in 1998 (DIW several editions) because of a sharp price difference between diesel-fuel and gasoline. Since 1985, technical progress in fuel efficiency has been partly offset by the increased size and power of new cars. The share of road transport in passenger and freight travel increased in the 1990s. Collective transport still plays an important role for commuting and other purposes in larger cities, but has lost share over the last 30 years. Because of these factors, the aggregate carbon emission intensity of travel rose slightly.

Since the late 1970s, growth in freight volume has been closely correlated with growth in industrial GDP, which is similar to the trend in other European countries. Trucks have gained in tonne-kilometres of travel at the expense of railroads. Germany's energy use for freight has been about average, relative to GDP, compared with its neighbours.

The German government has established a target of reducing the country's total CO_2 emissions to 25 percent below its 1990 emission level by 2005. Under the European commitment at Kyoto, Germany's target for 2010 amounts to a 21 percent reduction compared with 1990. In 1990, CO_2 emissions from transport amounted to almost 159 million

Table 4.1. Key Transport Statistics for Germany, 1997

Population (million)	82		
GDP per capita (US$ in PPP)	22,835	Cars (per 1,000 people)	504
Passenger transport activity (billion passenger-kilometres)	**901.4**	**Freight transport activity** (billion tonne-kilometres)	**450.1**
Passenger car	*82.1%*	*Road*	*67.1%*
Powered two-wheeler	*1.6%*	*Rail*	*16.2%*
Bus	*7.5%*	*Inland navigation*	*13.8%*
Urban rail	*1.6%*	*Oil pipeline*	*2.9%*
Railway	*7.1%*		
Transport final energy consumption (million tonnes of oil-equivalent)		**Transport CO_2 emissions** (millions tonnes)	
1990	**54.2**	1990	**158.7**
1997	**58.4**	1997	**173.4**
Rail	*4%*	*Rail*	*1%*
Road	*96%*	*Road*	*98%*
Inland navigation	*1%*	*Inland navigation*	*1%*

Source: *EU Transport in figures*, OECD, IEA, DIW 2000

Figure 4.1. Trends in CO_2 Emissions in Germany*, 1970-1996

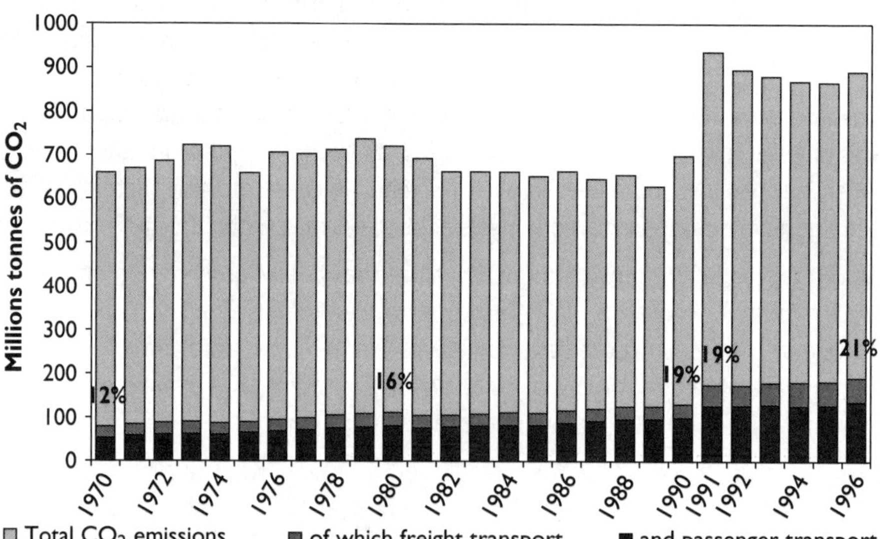

X% : Share of transport emissions in the total

* 1970-1990: West Germany only, Germany from 1991
Source: IEA/LBNL

tonnes, or about 15.6 percent of Germany's total CO_2 emissions. In 1997, they were up to 173.4 million tonnes (DIW 2000a, see Table 4.1), in contrast with the reduction in emissions from industry and the tertiary sector between 1990 and 1995, due to the economic downswing in the states of former East Germany.

Most forecasts for Germany suggest that present trends will raise CO_2 emissions more in the transport sector than in other sectors. Expectations are that until 2005, transport emissions will increase by 20 to 25 percent above 1990 (UBA 1997). Measures enacted by 1997 would bring 9 to 15 million tonnes of reduction.

Figure 4.1 shows trends in CO_2 emissions in Germany from 1970 to 1997.

Public Attitudes and Perceptions

Car manufacturing is an important part of the German industrial structure. Direct employment in car manufacturing accounts for 12 percent of the industrial work force (DIW 2000b). Directly or indirectly, every sixth job depends on this industry. Together with the national enthusiasm for cars, this makes the automobile "constituency" a strong economic force.

But the importance of this constituency may be changing. Surveys by the Federal Ministry for Environment have found that cars are considered by the population to be among the most significant polluters and a growing nuisance due to their noise and exhaust emissions. In larger cities, car-sharing networks, which allow people to have access to cars without owning one, are growing quickly, though from very modest beginnings. Cycling as a commuting mode is making a comeback in some areas. Public investments in transport generally have strong public support. More and more experts are calling for "integrated transport reform."

Other Influences and Trends

The full liberalisation of electricity markets that was enacted in Germany in 1999 could also affect local transport financing. There is some concern that decreasing revenues from municipal electricity production and distribution due to competition from large utilities will diminish municipal financial resources for public transport. Up to now, municipal transport that was not self-supporting was often cross-financed from electricity revenues within municipal organisations that combined energy, public transport and other municipal services. But European liberalisation of public transport markets is expected to reduce costs and subsidy requirements substantially through competition and privatisation.

Evolution of Transport-CO_2 Policy

Overview of Transport Policy

Federal policies on transportation are a shared responsibility of three ministries: the Ministry of Transport, Building and Housing (BMVBW); the Ministry of Environment (BMU); and the Ministry of Education, Science, and Technology (BMBF). The Transport Ministry has taken the lead in shaping policy by determining infrastructure policy, taxation and pricing in transport, and by leading the move to railway reform. BMU has no formal way to influence transport other than through emission legislation together with BMVBW. In general, the Environment Ministry and the subordinate Federal Environmental Agency do not place high importance on technological options for solving transport problems. In their research work, they emphasise traffic and urban planning, public transport and shifts in modal shares. BMBF is developing federal policy for science and technology. Together with BMVBW, BMBF emphasises technologies that influence systemic and organisational aspects of transport (telematics, traffic control and command systems) much more strongly than automotive technologies.

Providing for economic growth and mobility through efficient infrastructure and transportation systems has been, and still is, a prime objective of German transport policy at the federal level. This orientation has been reinforced since reunification. The fall of the Iron Curtain has meant more transit traffic through the old federal states, as well as more freight traffic in general. The Federal Traffic Infrastructure Plan, which was last updated in 1992, gives particular priority to the infrastructural needs in the new states. The Social Democratic government elected in 1998 has recognised the need for a new infrastructure plan. An emphasis on telematics to increase the capacity of the road network is a part of this strategy.

Compared with some other car-producing countries, Germany has in the past provided relatively little direct support in the form of R&D or demonstration projects for alternative vehicles and fuels (BMBF 1996). Partly in response to international competition and the feared loss of technological leadership, BMVBW announced in May 1998 a new strategic partnership, that brings together the federal government, car manufacturers, oil companies and a utility plan to commercialise the most promising sets of alternative engines, vehicles and fuels (BMVBW 1998b). The partnership seeks a fuel that will:

- Offer independence from petroleum.
- Be producible from renewable energy sources.
- Offer a reduction of CO_2 and pollutant emissions over its whole life-cycle.
- Work in a variety of propulsion systems, including internal combustion engines and fuel cells.

As of mid-2000, liquid natural gas, methanol, and liquid hydrogen had been retained as preferred options for further analysis (BMVBW 2000).

National transport policy continues to support the railway system by financing infrastructure, investing in vehicle stock, and making up for deficits in the budget of German Railways (Deutsche Bahn). The railway reform started in 1994 and is slowly converting the state-owned company into a private business. The reform aims to separate the infrastructure from the operation of three separated transport services: freight, short-distance passenger transport, and long-distance passenger transport, and to allow third parties to use the infrastructure. Competition is expected to improve productivity and traffic volume, and to achieve profitability. The divestiture of the still publicly – owned German Railways is expected by 2003. Whether these ambitious targets can be met without limiting rail service to the most profitable lines is still unclear. In early 2000, German Railways appeared to be contemplating a substantial reduction of working railway lines. This could imperil some of the CO_2 savings expected by the government from the railway reform and the related expansion of rail services (see Table 4.2).

With regard to local public transport, the federal government has shifted responsibility for planning and completing infrastructural projects to the states and municipalities. Most of the responsibilities, though not necessarily resources, for transport demand-management are now at this level of administration and therefore not subject to federal policy.

The Federal government's climate strategy is grounded in the exhaustive work of the Enquête-Commission of the German Bundestag (Enquête 1994) and has been set forth by the "CO_2-reduction" Interministerial Working Group (BMU 1994a and 1997b) that provided material for Germany's national reports to the UNFCCC in 1994 and 1997 (BMU 1994b, 1997b). A study commissioned in 1997 by the Federal Environmental Agency attempted to quantify the emission reduction for 2005 from measures already taken since 1990 and from those envisioned and under discussion (UBA 1997, BMU 1997b).

In the last German report to the UNFCCC (BMU 1997b), 25 measures were included for CO_2 emissions reduction in transportation. Apart from fuel and vehicle taxation measures and voluntary agreements with the car manufacturers, emphasis was put on a modal shift away from road and towards rail and public transport, as well as on technologies that can be used to optimise or regulate transport, such as telematics. Issues of land-use planning, traffic generation and urban transport are mentioned in the context of different research projects or are described as being in a planning stage.

While the voluntary agreements with car manufacturers were explicitly negotiated to respond to the increases in CO_2 emissions from private car use, other important measures have been implemented since 1990 that also affect CO_2 emissions. The most visible of these are:

- A voluntary agreement by German car manufacturers to reduce CO_2 emissions of new cars by 25 percent between 1990 and 2005; this is now superseded by the voluntary agreement on the European level, which mandates a 30 percent improvement by 2008 from the 1995 level.

- A significant increase in fuel taxes in 1991, 1994, and again under the eco-tax regime implemented in 1999, which promises further increases of DM 0.06 (€ 0.03) per litre each year until 2003.
- Environmentally-differentiated vehicle taxes, with the latest revision for freight in 1994 and for cars in 1997.
- The 1992 Federal Transport Infrastructure Plan, now enhanced by an infrastructure investment programme that tries to compensate for some of the shortfalls of the 1992 plan during the period between 1999 and 2002, when a new plan will be worked out.

A summary of all the transport-related measures in force that were quantified in terms of CO_2 reduction by the second national report to the UNFCCC (BMU 1997b, taken from UBA 1997) is given in Table 4.2.

Table 4.2. Measures Implemented since 1990 Quantified in Terms of Expected CO_2 Reductions (in megatonnes of CO_2)

	1990	*1995*	2000	2005
Base case (no measures)	*184.9*	*196.1*	231.0	236.0
Measure	Expected CO_2 reductions			
Amendment of the mineral-oil tax			3.5	5
Increasing use of local public transportation			3	3.4
Research programme on city traffic (FOPS)			0.5	1
Improving continuity of traffic flow			0.6	1.2
Freight centres			0.5	1
German Railways development concept for combined road and rail transport			0.3	1
Shifting international transit traffic from roads to railways and waterways			0.1	0.5
Sum*			6 to 10	9 to 15
*CO_2 emissions from new automobiles***			*3*	*7*

Source: UBA 1997, BMU 1997b

* Total of measures is not equal to the sum of quantities given for each measure due to interdependencies.

** In part overlapping with existing national voluntary agreements.

Note: CO_2 emission figures differ from data in Table 4.1 due to different sectoral delimitation.

Selected Transport-CO_2 Initiatives

Voluntary Agreements by the Car Manufacturers

Germany has a strong tradition of voluntary agreements by industry, and the government considers these agreements to be important elements in its national emission reduction strategy. For its part, the German auto industry has expressed concern in the past that fuel-efficiency regulation similar to the US Corporate Average Fuel Economy (CAFE) standards might be especially onerous for German manufacturers. They sell a disproportionate number of large, luxury cars that have higher fuel intensity than the average cars produced by other European manufacturers. So, in 1990 and again in 1995, the car manufacturers proposed to reduce the weighted average fuel consumption of their new cars by 25 percent between 1990 and 2005. The target value would be about 5.7 litres per 100 kilometres. In exchange for this commitment from industry, the federal government agreed not to impose new fuel-economy regulations. At the same time, the federal government supported the European Commission's efforts to negotiate a European-wide agreement (see Chapter 8). The German agreement played an important role in paving the way for the broader EU-wide agreement.

Fuel and Vehicle Taxation

Between 1990 and 1998, fuel taxes have been raised in two steps, leading to price increases of 30 percent for gasoline and 20 percent for diesel, thereby increasing the price differential between gasoline and diesel. The resulting incentive to use diesel cars was partly offset by increases in diesel vehicle taxes. Increased fuel taxation is considered important for CO_2 strategy because of its transport-restraining effects, which are expected to save an estimated 5 million tonnes of CO_2 by 2005. In 1999, the new government implemented an eco-tax regime that adds additional taxation on energy carriers. The revenues are supposed to finance a reduction of labour costs. The increases amount to DM 0.06 (€ 0.03) per litre of gasoline annually and they are planned to continue until 2003. This long-term approach is expected to reduce traffic and influence vehicle choice towards more fuel-efficient models.

Table 4.3. Yearly Fees for Motor Vehicles, DM per 100 cc displacement (and Euro equivalent)

Euro 3 or 4 (indicated level) or 90 g CO_2/km	Euro 2	Euro 1	Approved for Ozone Alert	Not Approved for Ozone Alert	All Others
Gasoline					
DM 10 (€ 5.11)	DM 12 (€ 6.14)	DM 13.20 (€ 6.75)	DM 21.60 (€ 11.05)	DM 33.20 (€ 16.97)	DM 41.60 (€ 21.27)
Diesel					
DM 27 (€ 13.80)	DM 29 (€ 14.83)	DM 37.10 (€ 18.97)	DM 45.50 (€ 23.26)	DM 57.10 (€ 29.19)	DM 65.50 (€ 33.49)

Note: the taxes on diesel motors are higher in part to offset the lower taxes on diesel fuel. In an ozone alert only cars meeting certain standards of pollution control can be driven.
Source: Bundestag 1996.

The previous government changed its vehicle taxation from a system solely based on engine displacement towards a system also based on emissions. While taxation of trucks above 3.5 tonnes was introduced in 1994 and includes pollutant and noise-emission features, the emission-differentiation component of car taxation introduced in 1997 also includes some differentiation with regard to CO_2 emissions (see Table 4.3). Cars with CO_2 emissions below 90 grams per kilometre – 3.8 litres of gasoline or 3.4 litres of diesel per 100 kilometres – are included in the lowest vehicle taxation class even if they fail to comply with the European pollutant-emission regulation standards, Euro 3 or Euro 4. "Euro 3 and 4" cars introduced before the date at which each regulation became active (2000 and 2005, respectively), and cars with CO_2 emissions lower than 90 grams per kilometre, are granted vehicle-tax exemption up to a cumulative amount of DM1,000 (€ 511). This measure is intended to be a start towards a more differentiated taxation system, one in which the tax changes according to certain performance indicators, such as fuel consumption. The objective is clear: the early introduction of low-polluting or highly efficient vehicles. A similar approach was taken during the late 1980s to stimulate the introduction of vehicles equipped with catalysers.

In its comprehensive review of revised vehicle taxation, DIW, a major economic research institute, brought up several arguments suggesting that, although the approach seems to go in the right direction, its effects might fall short of expectations (DIW 1997):

- Giving an advantage, or even exemption, to cars equipped with catalysers did not trigger a particularly quick introduction of the new technology.
- As to CO_2 differentiation, no cars yet on the market could benefit from the exemption and the lower vehicle taxation.
- Three-litre-per-100 km cars will have engines with relatively low displacement volume. Because the vehicle tax is still proportional to motor displacement, the fiscal advantage of a standard-efficiency/low-displacement car over a high-efficiency/low-displacement car is marginal. A progressive tax based on fuel consumption or emissions, instead of engine size, could have a more distinct effect.

The 1997 changes in vehicle taxation grew out of lengthy deliberations, and they represent a first step in integrating CO_2 emission features into the fiscal framework. The effects will be reviewed in 2002, and the question of switching the emphasis from vehicle taxation to fuel taxation will receive particular consideration. Because revenues from vehicle taxation go to the states and revenues from fuel taxation go to the federal government, a complete transfer of taxation from vehicles to fuel seems unlikely. In fact, there are good reasons to retain differentiated vehicle taxation because it could be an effective tool for influencing the composition of the overall vehicle stock. Gearing it towards fuel efficiency or emissions, rather than size, would be an obvious way to make vehicle taxation more effective. By 2002, higher-efficiency vehicles ("three-litre cars") will be available, and it may be that fiscal support, through a thoroughly restructured fuel and vehicle tax system, will prove necessary to make such cars commercially attractive.

Federal Transportation Infrastructure Plan

In 1992, the Federal Transportation Infrastructure Plan (BMV 1992) established the framework for infrastructure investments in all transport modes up to 2010, with particular emphasis on the new situation after the reunification of Germany and subsequent fall of communist governments in most of eastern Europe. The plan also took into account the increasing integration of the European internal market and expected increases in transport demand. Concern about growing CO_2 emissions from transportation were explicitly addressed and reflected in a shift in the structure of the investments. The share of rail infrastructure and local public transport increased, and the share of road infrastructure decreased. Since 1990, 43 percent of all federal investments in transport infrastructure have been made in the new states, of which more than half was spent on railways, almost a third on inter-city roads and motorways, and about 20 percent on municipal roads and public transport (BMVBW 1998). The Enquête Commission estimated that the expected shift towards rail transport would reduce CO_2 emissions by 2 to 3 percent by 2005, relative to a continuation of former trends (Enquête 1994).

There is much debate about whether these expectations will be fulfilled. Due to budget constraints after reunification, not all projects developed as initially envisaged, and this could delay some of the expected effects. It is also hard to assess how much the change in expenditure really reflects a conscious redirection from road to rail infrastructure. It is the first plan that includes the new states, and a high share in expenditure was dictated by the urgent requirements there. The Enquête Commission recommended speeding up and further emphasising intermodality and railway system improvements (Enquête 1994). It is still not clear whether the investments in rail travel will suffice to induce modal shifts, unless it is supported by a consistent policy that restrains growth in road transportation. Finally, there is ongoing debate on the impact of improved road infrastructure. One side argues that it will foster smooth traffic flow, thus keeping fuel consumption per kilometre travelled lower than in more congested situation. The other side emphasises that further increases in road infrastructure will increase road transport activity in the longer term. The philosophy of the current infrastructure plan seems to follow the former argument, for it states that transport policy should ensure that "infrastructural bottlenecks" do not hinder economic growth. The government expects that emissions will be reduced because excessive fuel consumption in congested traffic will be avoided (BMU 1997a). As a matter of fact, the framework projects considerable growth in road infrastructure and does not explicitly address the problems of demand generation and land-use. Various research projects are underway that include these aspects, as well as a measure called "introduction of traffic-effect studies" that could be integrated into infrastructure planning (BMU 1997b).

The Social Democrats and Green Party elected in 1998 have stressed the need for a new infrastructure plan. The underlying traffic projections for the previous plan, still in effect, proved to be too low, and the whole plan was underfinanced. The future plan, which is yet to be developed, is supposed to focus on measures that would increase rail's share in freight, increase integration of the different transport systems (road, rail, and waterways) and improve existing infrastructure rather than expand infrastructure for rail and road (BMVBW 1999).

Further Measures

To support a modal shift towards railways and combined freight transport, other measures have been implemented since 1990 that are expected to have a CO_2-reducing effect. Several research projects on city and land-use planning and mobility, although they will have no immediate effect, may influence decision-making and regulation in the longer run. Information technologies are expected to improve the continuity of traffic flow and reduce and regulate traffic, thus reducing CO_2 emissions. A quantification of these measures is not possible, and the direct effects on emissions are expected to be low.

Since 1995, together with the Benelux countries and Denmark, Germany has imposed a yearly road-use fee on heavy trucks. This ensures that foreign trucks contribute to domestic roadway infrastructure costs. As a side effect, expansion of trucking on motorways might be limited to some extent. Because it is a yearly fee based on truck size and not on distance travelled or emissions, this measure probably has only a weak effect on CO_2 emissions. Yet the measure is a first step towards future and more coherent full-cost pricing of road freight with the help of electronic road pricing. It has been recognised that a kilometre charge for trucks, originally adopted as a means to internalise infrastructure cost, could help to balance the competitiveness of different freight modes to the benefit of rail and waterway transport. Such a scheme, under consideration for 2002-03, would replace the annual fee implemented in 1995.

The government has recognised that measures now in force will not achieve a substantial reduction of CO_2 emissions from transport. The study group (UBA 1997) proposes a number of further measures to compensate for expected emission increases in their baseline projections. These measures include speed limits, stringent fleet-efficiency standards, strong increases in fuel price, road pricing, and general training in energy-efficient driving. The study group estimates that these measures will achieve a CO_2 reduction of more than 10 megatons – far more than the measures implemented to date. While the study group believes that such a package of measures could stabilise emissions, they make clear that such a "heavy-handed" approach is politically unfeasible. The group offers an alternative scheme, called "programme of measures for passenger and freight transport." This programme, which assumes a societal consensus on the risks posed by increasing transport and unchanged transport patterns, would restrain road use and foster alternative travel and freight modes through a co-ordinated basket of measures, including regulation, pricing, investment in infrastructure and land-use policies. While such an approach needs to be carefully co-ordinated, the proponents claim that it will produce higher CO_2 reductions, especially in the longer term. The study argues that current measures are not well enough integrated and that no long-term strategy beyond 2010 is yet visible. The government seems to have recognised some of its predecessor's shortfalls. Yet, new measures, other than additional fuel taxation under the eco-tax scheme, are yet to be developed.

Conclusions

Even if the German government's strategy is somewhat dispersed and many of its measures may need further definition and integration, there are indications that measures already implemented will reduce the growth of emissions. These include the voluntary agreements by car manufacturers, railway reform, and steps towards a differentiated taxation and pricing policy.

The current set of measures, however, does not yet amount to a coherent longer-term strategy for transportation beyond 2010. What seems necessary is an integrated set of measures aimed at a common target. The draft environmental programme (BMU 1997a), which contains specific targets for transportation, could provide a focal point and a yardstick for measuring progress towards a sustainable transportation system. Transport policies at the federal, state, regional and municipal levels need to be integrated and directed towards these targets. Taxation and pricing of car use is one area where progress has already been made, but where a more coherent approach, extending from federal fuel taxation to municipal parking pricing, could reinforce the effectiveness of other measures.

As pointed out before, the expectations put on the voluntary agreements will probably be met only if complementary measures for market deployment of highly efficient cars are taken, particularly fuel tax increases. Similarly, modal shifts cannot be expected to arise from the improvement of rail services alone, but should be supported by road-restraining measures, such as the kilometre charge for trucks. Finally, research on traffic generation, land use, and urban planning should be taken into account and integrated into federal policy, particularly on infrastructure. Transport policies by the states and municipalities to limit transport must be further supported and taken into account when measures on the federal level are developed. Technologies, like telematics applications, should be carefully assessed with an eye to their possible adverse effects, and care should be taken on how to integrate them into an environmentally consistent transport policy.

CHAPTER 5: THE NETHERLANDS

Background

Trends in Transport and CO_2 Emissions

Transportation in the Netherlands is typical of the rest of Europe in some respects, but unusual in others. Although car ownership and per-capita usage is below average, total domestic passenger travel is about the same as elsewhere. Low car ownership is offset by the highest bicycling and walking rates among the wealthy countries in western Europe. High purchase taxes on cars play a role in discouraging ownership, but high population density, good inter-city rail service, and a system of bike paths both within and between towns also helps to reduce car travel.

In terms of fuel mix, the Dutch car fleet is unique in western Europe: Liquefied Petroleum Gas (LPG) competes with gasoline and diesel fuels. Gasoline is used in roughly 70 percent of all vehicles and LPG and diesel share the rest.

The Netherlands is a densely populated country. Traffic is congested both within and between cities. Careful planning of building locations and transport infrastructure has always been a high priority, with significant emphasis placed on promoting alternatives to automobile travel. Nonetheless, because of the Netherlands' position on the sea, truck traffic continues to be quite high. The government recognises the enormous burden placed on the Dutch environment by transit traffic. At the same time, the mobility of goods is important. Rotterdam is the world's leading port in terms of total cargo handled. Geographically, the Netherlands is a corridor from the North Sea towards Germany and central Europe. More than half of the Dutch GDP comes from international trade.

Figure 5.1 shows trends in Dutch CO_2 emissions from 1970 to 1997.

Table 5.1. Key Transport Statistics for the Netherlands, 1997

Population (million)	15.6		
GDP per capita (US$ in PPP)	23,082	Cars (per 1,000 people)	372
Passenger transport activity (billion passenger-kilometres)	**198.3**	**Freight transport activity** (billion tonne-kilometres)	**95.4**
Passenger car	*76%*	*Road*	*47%*
Powered two-wheeler	*1%*	*Rail*	*4%*
Bus	*7%*	*Inland navigation*	*43%*
Urban rail	*1%*	*Oil pipeline*	*6%*
Railway	*7%*		
Bicycle	*1%*		
Transport final energy consumption (million tonnes of oil-equivalent)		**Transport CO_2 emissions** (millions tonnes)	
1990	**8.9**	1990	**25.9**
1997	**10.7**	1997	**31.2**
Rail	*2%*	*Rail*	*0%*
Road	*91%*	*Road*	*93%*
Inland navigation	*7%*	*Inland navigation*	*7%*

Source: *EU Transport in figures*, OECD, IEA

Figure 5.1. Trends in CO_2 Emissions in the Netherlands*, 1970-1997

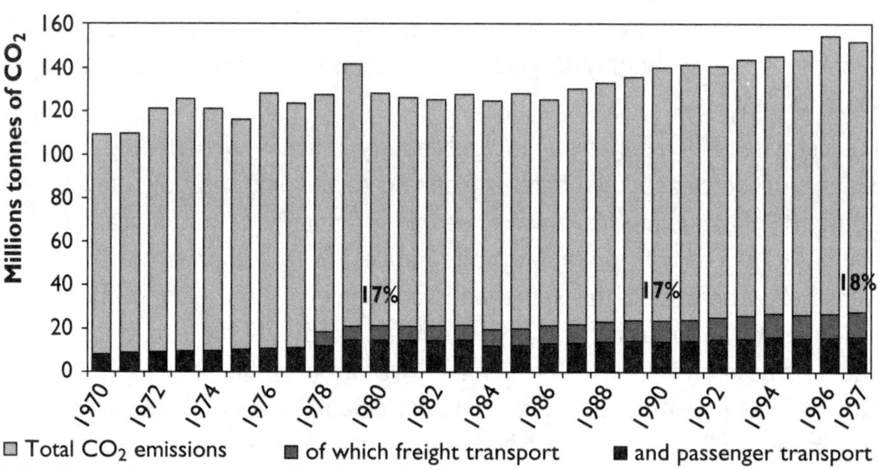

X% : Share of transport emissions in the total

* No data on energy consumption of trucks is available before 1978, therefore freight energy consumption figures are shown from 1978 onwards.
Source: IEA/LBNL

Public Attitudes and Perceptions

The Dutch accept strong government intervention in all parts of the economy, including transport. Both regulations and fiscal stimuli are seen as legitimate elements of policy in general, and as important elements of transport and environmental policy in particular.

A tradition of bicycling and walking, along with good systems of public transit in most cities, makes it easy for the average citizen to accept changes that serve to reinforce these modes.

While the Dutch follow EU standards for improvements in air pollution, they also aim to reduce CO_2 emissions from transportation through policies that seek behavioural change. Stricter enforcement of speed limits, the encouragement of more careful driving (including installation of econometers, on-board computers and cruise control in new cars), and even limitation of new-car power are currently under consideration. The exemption of econometers (which show drivers how much fuel a car is using instantaneously) and other related equipment from the purchase tax has been in place since May 2000. The Netherlands Agency for Energy and the Environment (NOVEM) has launched an educational campaign to promote lower speeds and more fuel-economic driving and to discourage unnecessary driving. The Dutch authorities have submitted a proposal to the European Union for labelling car fuel economy based on a car's fuel economy within its overall car class.

At the same time, there is a still strong pro-car sentiment in the country. In 1989, a government fell over a new scheme for travel-cost allowances for commuters. Although measures to encourage the purchase of certain energy efficiency equipment were eventually put in place, Parliament had voted against them in December 1997. Road pricing as a mean to relieve congestion has been repeatedly proposed in the past, but is not yet politically acceptable.

Role and Position of Industry

The Netherlands does not have a real domestic car industry. The few car factories located in the country are all owned by foreign companies. This may make it easier, politically, for the Netherlands to maintain high taxes on vehicle purchases. Dutch authorities are largely dependent upon international efforts to develop and market low-pollution, fuel-efficient vehicles. Nonetheless, the government has carried out a series of evaluations of technology and estimated potential reductions in fuel consumption (VROM 1996).

However, there *is* a car association (Nederlandse Vereniging de Rijwiel- en Automobiel-Industrie, RAI), which has opposed some important government initiatives. Raising diesel prices has met with stiff resistance from the trucking industry.

Shell International, which has part of its home in the Netherlands, has recently taken several highly publicised steps towards "sustainable energy". This step suggests that the Dutch Oil industry, which had previously been silent on the subject, now recognises the need to cut carbon emissions.

Other Influences and Trends

Perhaps the single most important influence on transport policy in the Netherlands is the crowded nature of the country. Local governments are taking into account land-use patterns and life quality by carving out large car-free zones in cities, which are also criss-crossed with bicycle paths.

In recent years, much road congestion has been between cities rather than within cities. An understanding public may eventually accept strong measures to control the flow of vehicles, local pollution, and noise traffic.

Evolution of Transport-CO_2 Policy

The Netherlands recognises that transport, energy, environment and physical planning issues often overlap. Several ministries – the Ministry of Transport (Ministerie van Verkeer en Waterstaat, V&W), the Ministry of Housing, Spatial Planning and the Environment (Ministerie van Volkshuisvesting, Ruimtelijke Ordening en Milieubeheer, VROM), and the

Ministry of Economic Affairs – all have certain responsibilities related to transport as well as CO_2-emission reductions. Discussions of CO_2 appear routinely in the relevant policy documents of each ministry. As a result, CO_2 reduction receives little special treatment on its own beyond general statements of government goals or submissions to the UNFCCC. CO_2 policies are most often formulated by the authorities responsible for the activities that cause the emissions. There appears to be good co-ordination among authorities. The fact that housing, environment, and spatial planning are in the same ministry itself gives a strong boost to this co-operation.

The Netherlands' CO_2 policy is based on a series of *National Environment Policy Plans* (Nationaal Milieubeleidsplan) that provide targets for environmental improvement and outline policy action in each major sector. Alongside these are a series of specific policy plans – on physical planning, transport, and climate. These plans have set quantitative goals related to traffic, energy use, local air pollution, safety and CO_2 emissions.

Progress towards the goals laid out in these plans is generally tracked with the aid of various indicators. Transport indicators include indices of traffic volumes, expected fleet-wide values for CO_2 (and other) emissions per kilometre and so forth. Guided in part by a car-use panel and a mobility survey held continuously since 1979, the government has monitored changes in transport activity, energy use, emissions, congestion, noise, and other factors. Results of government policies are evaluated in a series of outlooks ("Verkenning", meaning literally "exploration") reports. Thus, the government has made a strong effort to quantify its goals and progress, and it has been frank in developing and publicising indicators as signposts of success... or failure.

In the early 1990s, the *Second Transport Structural Plan* (SVV-II) (V&W 1990) and the *Fourth Report on Physical Planning* (Vierde Nota Ruimtelijke Ordening Extra, VROM 1993) laid the groundwork for many CO_2 policies. SVV-II focused on the theme of managed mobility. It called for tying every major residential development to high-grade public transport and controlling parking, particularly employer-provided parking.

Charging for road use at congested periods was discussed at the time, but not implemented. Raising diesel prices met with resistance from truckers and trucker unions, so little progress has been made to date. Doubling peak-hour collective transport capacity by 2010 was another important goal, as was expanding the trunk road, rail, and waterway networks.

The Amended Fourth National Plan on Physical Planning (VROM 1990) emphasised the importance of land-use planning in limiting the growth of car use. It proposed that municipal decision-making be based on the proximity of housing, employment, and recreation areas and the ability of people to travel between them by modes other than personal road vehicles.

The 1993 Nationale Milieuverkenning 3 (MV3, Third National Environment Policy Plan) (RIVM 1993a and 1993b) called for attacking environmental problems from transport and restraining emissions of CO_2 through a variety of tax, price, and regulatory measures. Goals for the year 2000 included limiting the increase in car distance driven to 135 percent of the 1986 amount and holding CO_2 emissions constant at the 1986 rate through improvements in fuel economy. The goal was 26 percent less fuel per kilometre. For 2010, the plan called for reducing emissions by 10 percent from the 1986 values. Only some of the MV3 goals have been matched with concrete legislation, and not all of the goals are being met.

The Second Memorandum on Climate Change (VROM 1996) proposed an emission ceiling for road transport of 23 megatons of CO_2 in 2000, with a 10 percent reduction by 2010. The report frankly acknowledged the difficulty of achieving these goals, and recent developments have already overshot the 2000 target.

A report on Accessibility (V&W 1996) emphasised physical location as an important element in reducing passenger traffic. This document addressed the obstacles, particularly growing congestion and the rapid rise in road freight, to meet the goals in the *Second Transport Structural Plan* (SVV-II). More traffic management and collective transport measures were proposed, and the Netherlands' two main ports, Rotterdam Seaport and Schiphol Airport, were singled out as needing improved accessibility. Road pricing was also suggested, as were stronger parking measures. At the same time, *Transport in Balance* (V&W 1996) addressed rising traffic, congestion, and emissions from freight and set specific goals for 2010. These include a 5 percent decrease in long-distance domestic road freight, 10 percent less international road freight, and a 40 percent reduction in local traffic tonne-kilometres by road.

In 1998, the fourth National Environment Policy Plan (MV4) (RIVM 1998) renewed a number of important goals, including carbon emission restraint (anticipating the targets that would be discussed later that year in Kyoto). It also showed that present trends and policies would not provide significant restraint. MV4 suggests that stronger measures are required to reduce the social costs of energy use and restrain traffic in general. These include more tax breaks for using public transport and restricting the tax deductions for using private cars to get to work. It also recommends stimulating telecommuting, temporarily freeing bio-ethanol from excise taxes for experimental projects, and raising excise taxes on other fuels according to environmental damages. These are not CO_2-specific measures, but they will all lead to restrain CO_2 emissions.

MV4 also sets bold goals for reducing the use of light truck diesel fuel and taxi fuel in favour of LPG. The measures are primarily aimed at reducing emissions of NO_x, PM10, and VOC, but should also have a beneficial effect on CO_2. One of the goals is to increase the share of LPG in the transport sector, to reach a level in 2010 that would be equivalent to the present diesel share in passenger car energy consumption. Since the government is willing to change new-vehicle and yearly vehicle taxes, as well as fuel taxes, this is likely to be achieved. But its impact on CO_2 emissions will be small, particularly as some kinds of LPG vehicles are actually *more* fuel intensive than their gasoline (or diesel) counterparts.

In 1999, VROM issued a document on climate policy implementation announcing the measures and policies required to meet the Netherlands' share of the European Union's emission reduction obligation under the Kyoto Protocol (VROM 1999). The agreed upon target for the Netherlands is a 6 percent reduction. This time the Dutch policy-makers did not choose to set targets for different economic sectors. For transport they announced the following measures and instruments:

- Application of the EU voluntary agreement for more fuel-efficient cars.
- A "feebate" system for car purchase tax.
- Better enforcement of current speed limits.
- Road pricing.
- Fiscal measures to discourage car use in commuting.
- Increased tire pressure to reduce fuel use.
- Promotion of the use of econometers, cruise control, and on-board computers in cars.
- Projects aimed at more fuel efficiency.

Aside from the EU voluntary agreement, however, only the measure promoting the use of econometers, cruise control and on-board computers in cars is in force now. According to this new plan, transport emission reductions would amount to 2.2 to 2.9 million tonnes in 2010, compared to 37.9 million tonnes if no measures were taken. Of the 6 percent economy-wide reductions of 25 million tonnes, transport will contribute a share of 9 to 11 percent (VROM 1999). In a research evaluation of the 1999 VROM plan (ECN/RIVM 1999), the estimated reduction in emissions is less than what is suggested in the policy document. The policy set for transport is estimated to result in a total reduction of 1.3 to 2 million tonnes of greenhouse gas emissions by 2010, compared with the business-as-usual scenario.

Selected Transport-CO_2 Initiatives

The Dutch approach rests on many measures related to pricing and behaviour. Since June 2000, fuel taxes have risen slightly. More important, tax shifts are underway to increase the variable costs and lower the fixed costs of driving. Road pricing and other measures that affect driver behaviour are planned or still under discussion. New-car fuel-economy labelling (in the form of a proposal to the European Union) stricter speed limits, and various ways of instrumenting cars for greater fuel economy are still under discussion.

Van Wee and Annema (1999) estimated that the base case emissions from transport in 2010 would rise to 37.8 million tonnes of CO_2, of which 17.8 million tonnes would arise from cars and 16.0 million tonnes from other road transport.

Variabilisation of Vehicle Costs and Higher Fuel Taxes

The principle of variabilisation recognises that most of the social costs of transport are related to the use of vehicles and the system, not to ownership per se. Variabilisation seeks to shift the fiscal burden of transport from the fixed costs placed on vehicle or system ownership to taxes on use. An effort is made to differentiate both fixed and variable costs according to "environmental friendliness."

A significant step towards variabilisation was taken in July 1997, when fuel taxes were raised and annual vehicle taxes were lowered. Eventually, higher fuel taxes and environmentally differentiated new-vehicle and yearly taxes are expected to go into effect as the political climate permits. But the extent to which fuel and vehicle taxes can be adjusted is limited to some extent by pricing policies in neighbouring countries. Since the United Kingdom has raised diesel taxes to near parity with those on gasoline, and France has begun to do so, more movement in this direction can be expected in the Netherlands. At the same time, fixed costs for vehicles will be more sharply differentiated, so that diesel vehicles will cost more to buy than they do today. The government proposes to lower new-car purchase taxes for the least fuel-intensive models in each car class.

The most obvious policies are higher taxes on diesel and LPG as road fuels, since low LPG and diesel fuel prices, as well as light company-car taxation, currently encourage driving of diesel and LPG cars. Despite the higher purchase taxes on diesel and LPG vehicles, long-distance drivers have tended to pick them in the hope of recouping their higher fixed costs through lower running costs. But diesel in particular has been identified by the Dutch authorities as causing problems for both smog and particulates. The Dutch government will try to reverse the trend in diesel use by encouraging LPG, natural gas, or dimethyl-ether (DME) for buses and LPG, instead of diesel, for light trucks and vans. These policies will be buttressed both by pricing changes and by shifts in new-vehicle and yearly vehicle taxes according to fuel consumption. In general, the government wants the fuel cost of driving at least to remain constant in real terms and so proposes to increase prices at about the same rate at which (they hope) fuel economy of new cars improves.

Over the longer term, variable costs will rise and fixed costs will fall. This was made clear in the 1 July 1997 shifts in diesel and LPG taxes. The changes in fuel taxes were small compared to prices at the time, as shown in Table 5.2. But fuel-tax increases on gasoline and diesel were proposed in the climate policy implementation plan (VROM 1999) – by € 0.22 per litre on 1 January 1999, and by the same amount on 1 January 2003. Van Wee and Annema (1999) estimated that these two price increases alone would reduce emissions from the base value in 2010 by 4.2 million tonnes of CO_2, about 12 percent of the expected level in their base case.

These new taxes and the first step towards tax variabilisation went into effect on 1 July 1997. Many of the transport measures have been put into

Table 5.2. Variabilisation and Green Measures for Dutch Cars, 1 July 1997

In Dutch Guilders (f) (and euros)	Gasoline	Diesel	LPG
Fuel Increase* excluding VAT	+ f 0.11 (€ 0.05)	f 0.05 (€ 0.02)	+ f 8 (€ 0.04)
1997 price per litre before increase	f 2.18 (€ 0.99)	f 1.50 (€ 0.68)	f 0.8 (€ 0.36)
Yearly Registration Fee (Motor Rijtuigen Belasting, MRB)			
Reduction of Car MRB	f 140 (€ 63)	f 107 (€ 48)	F 220 (€ 100)
Reduction of Bus MRB	–	45.8 percent	100 percent**
Reduction of Vans MRB	f 178 (€ 81)	f 178 (€ 81)	f 178 (€ 81)
Zero Taxation on Electric Cars			
Reduction in LPG Car MRB with G3 emission control			f 470 (€ 213)

* Greater increases are foreseen for the future. These occurred in 1999, 2000, and are scheduled for 2003.
** Includes natural gas buses.
Source: VROM.

place, but many goals related to CO_2 in transportation have not yet been met, a theme echoed in VM3. As noted by van Wee and Annema (1999), the effects of the small changes so far were still too weak to achieve significant steps towards the present goals.

Land-Use Planning

The Netherlands recognises the importance of administrative measures, particularly those related to land-use planning and information (V&W 1996). Reducing employer-provided parking and making good public transportation available at every major new residential development are measures mentioned by all authorities. There has also been discussion of regulating the location of new places of employment and residence according to their accessibility to collective transport (van Wee and van der Hoorn 1996). Because this approach exercises push (through regulations and fiscal measures) and pull (through greater provision of alternatives to car use), it is likely to have a noticeable long-term impact on total traffic.

Most authorities stress the importance of land-use planning, as set forth in the Fourth National Plan on Structural Planning (VROM 1993). Various studies show that modal choice is dependent upon density and proximity. The "ABC" approach (van Wee and van der Hoorn 1996) grades locations according to their accessibility by walking/biking or collective modes.

Other Measures

The Ministry of Environment emphasises stricter speed limit enforcement. Speed limit enforcement per se is given a prominent position in the Fourth National Environmental Policy Plan (VROM 1998). Econometers, cruise control and on-board computer monitoring of engine performance, all of which would encourage more energy-efficient driving, were exempted from purchase tax effective on 1 May 2000

Road pricing is scheduled to start in 2001. Authorities believe that the technologies needed to allow remote, anonymous charging are well-enough developed to support a full system. They now plan to implement such a system in highly congested intercity regions. Because road pricing will be limited to a few of the most congested intercity motorways, its effect on total distance driven, and thus on CO_2 emissions, will be rather small. Van Wee and Annema (1999) put it at less than 1 percent of total transport emissions by 2010.

The government also emphasises improving the freight system (V&W 1996). It has set out explicit goals for modal shifts and overall growth (of road, water and rail transport), for intermodal facilities, for varying costs, and for infrastructure improvement. If the targeted modal shift and volume changes come about, these alone could reduce freight carbon emissions by approximately 10 percent from their expected values by 2010 and beyond (VROM 1999).

Conclusions

Several factors contribute to the potential success of the Dutch effort. The most important of these is a longstanding acceptance of government regulations and fiscal stimuli as legitimate tools for policy in general, and important elements of transport and environmental policy in particular. There is a clear tradition of inter-ministerial co-operation, as evidenced by the way in which energy, environment, transport, infrastructure, and housing fit together in policy documents. The fact that housing, environment, and physical spatial planning are in the same ministry gives a strong boost to this co-operation. A tradition of bicycling and walking, and good public transport makes it possible for the average citizen to accept changes that favour these modes. Heavy traffic makes the acceptance of road pricing likely, at least in the most congested corridors.

Strong growth has occurred, nevertheless, in vehicle travel, both passenger and freight, and further measures to meet future CO_2-emission reduction and other environmental goals are needed.

One key lesson for policy-makers is that it is possible to impose CO_2 restraints through policies aimed at local transport and environmental problems, but not tied directly to CO_2. In its national plans, the government explicitly plays down CO_2, compared with other problems facing transport. Thus, most measures are justified because of other concerns related to transport, such as safety, air pollution or congestion. Even if reducing CO_2 were not a goal, transport reform in the Netherlands would still reduce CO_2 emissions by 2010 by some 10 to 15 percent compared to what they would have been without these measures.

CHAPTER 6: SWEDEN

Background

Trends in Transport and CO_2 Emissions

With two domestic car companies, Sweden is well motorised. Per capita car use is higher than in countries with similar GDP, reflecting its sparsely populated territory. Growth ownership has slowed recently, in part because of approaching saturation, giving Sweden one of the oldest car fleets in Europe. Sweden's cars are also the heaviest in Europe, and they are highly powered as well. Not surprisingly, fuel intensity has been higher than the European average. New-car fuel economy was affected by a government voluntary agreement (VA) with manufacturers and importers, which took effect in 1978. The VA's target of 8.5 litres per 100 kilometres was surpassed during the 1982-92 period (with rates between 8.2 and 8.3 litres per 100 kilometres). High rail use, average bus use, and a large amount of domestic air travel, along with heavy car use, have made Sweden one of the heaviest travelling countries in Europe.

Because Sweden is a major exporter of raw materials, particularly iron ore and forest products, and because of long distances from production areas to ports, freight transport is very important to the Swedish economy. The ratio of total domestic freight to GDP is high, though the share of trucks is low by European standards. At the same time, fuel use per tonne-kilometre for trucking is among the lowest in the IEA, mainly because Sweden permits very large trucks (up to 60 tonnes). While the empty back-haul rate has been rising slowly, it is still only around 30 percent for trucking for hire, lower than in the Netherlands or Denmark. The use of smaller trucks is increasing and accounts for most of the rise in the fuel intensity of trucking in Sweden. Although Sweden's economy is freight-intensive, the combination of a low share of trucks and relatively low fuel

Table 6.1. Key Transport Statistics for Sweden, 1997

Population (million)	**8.8**		
GDP per capita (US$ in PPP)	**21,213**	Cars (per 1,000 people)	**419**
Passenger transport activity (billion passenger-kilometres)	**112.1**	**Freight transport activity** (billion tonne-kilometres)	**52.2**
Passenger car	*83%*	*Road*	*63%*
Powered two-wheeler	*1%*	*Rail*	*17%*
Bus	*8%*		
Urban rail	*1%*		
Railway	*6%*		
Transport final energy consumption (million tonnes of oil-equivalent)		**Transport CO_2 emissions** (millions tonnes)	
1990	**6.6**	1990	**18.7**
1997	**6.9**	1997	**19.7**
Rail	*4%*	*Rail*	*1%*
Road	*94%*	*Road*	*97%*
Inland navigation	*2%*	*Inland navigation*	*2%*

Source: *EU Transport in figures*, OECD, IEA

Figure 6.1. CO_2 Emissions in Sweden, 1970-1997

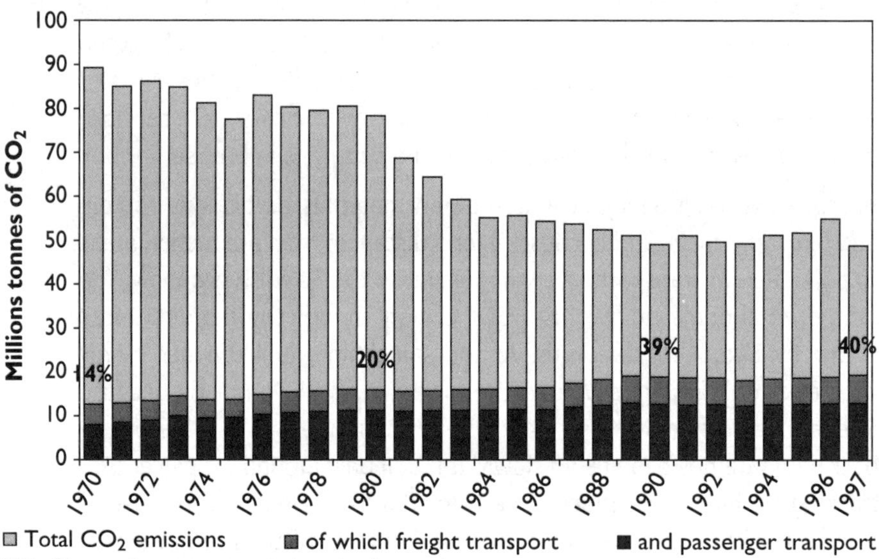

X% : Share of transport emissions in the total

Source: IEA/LBNL

intensity of trucking, somehow leads to freight fuel relative to GDP that is about average for European countries.

Figure 6.1 shows trends in CO_2 Emissions in Sweden from 1970 to 1997.

Public Attitudes and Perceptions

Sweden has always been a "driver-friendly" country. With relatively low taxes on cars and driving, and good road infrastructure, Sweden's public has for decades enjoyed the highest mobility (in passenger-kilometres per capita) of any country in Europe. The Bruntland Commission report in 1987 sparked a long public debate over nuclear power, which in turn lead to a debate over the environment and sustainable development. The public entered the transport debate in 1989 and 1990 with gusto. Citizens now seem prepared to deal with the issue of transport and CO_2, even after a press and parliamentary debate has continued now for 10 years. The tax revolt that marked many Scandinavian elections in the 1980s has also faded, as has the popularity of some groups who opposed major transport reforms.

Role and Position of the Industry

With two car and truck companies, in a country with only 8.5 million people, producing between 250,000 and 400,000 cars per year, Sweden is one of the world's largest producers per capita of motor vehicles. While over ambitious environmental goals might endanger this domestic industry, initiatives from *within* the industry may have a profound positive impact on restraining emissions and be profitable to the companies.

Volvo's pledge of a 25 percent reduction in the fuel intensity of new cars by 2005 surprised many, because it was made unilaterally and, at least from the general public, was not expected. According to Volvo, meeting this pledge will involve some down-sizing and weight reductions, the introduction of some diesel motors and the use of other fuels (Wallman 1996). Because Volvo sells mostly heavy cars, there is room for some down-sizing. Senior authorities at Volvo believe they will be able to lower their present sales-weighted fuel average from roughly 9.5 litres per 100 kilometres to 7.5 litres. However, the officials do not believe they could reach the 5 litre per 100 kilometre target discussed in the European-wide ACEA Voluntary Agreement or even the 6.25 litre per 100 kilometre target. But now that Volvo has been acquired by Ford, a move which raises the possibility of including smaller cars under the Volvo name, and General Motors has raised its stake in Saab, the Swedish car industry may be set to sell smaller cars.

While Volvo has positioned itself as a leading environmentally conscious company, there is an active "automobile lobby" in Sweden, as in other countries with auto manufacturers. Among the key actors, Volvo seems ready to go halfway towards the Communications Committee's (KOMKOM) fuel economy goal of a 15 percent reduction in total fuel use by 2020. This means a 25 percent reduction in fuel use per kilometre, enough to offset some of the expected growth in traffic. But the net result would still be an increase in total road fuel use. The Swedish Automobile Association is prepared to co-operate, too, but is understandably cautious about making any commitments.

Other influences and Trends

Sweden is a sparsely populated country, with a few large cities that have taken the lead in forging local clean-air policy in all sectors. This geographical fact has created two distinct political groups: city dwellers that must live with air pollution and traffic congestion in the cities, and the rest of the population. Sweden has high driving distances per car. The government and drivers have always favoured heavy, "safe" cars, and the two Swedish auto manufacturers have become world leaders in producing such cars. The large cars have also been supported by company-car policies, as will be noted below. Thus, reducing carbon emissions from transportation may seem like a major challenge in view of Sweden's traditions of large cars and open roads.

Evolution of Transport-CO_2 Policy

Behind the authorities' determination to restrain or reduce CO_2 is the fact that Sweden has achieved one of the strongest air-quality improvements in all sectors of any country in Europe. This started with the use of district heating to reduce heavy oil burning from uncontrolled boilers in cities. More recently, Sweden has become a European leader in technology and regulations to reduce emissions from motor vehicles. This resulted in the early introduction of lead-free fuel and of very low-sulphur diesel fuel (ethanol for busses), supported in part by differentiated taxation on fuels according to quality and likely pollution. Fuels and new cars are taxed according to their "environmental classes." Taxes on new cars with very good pollution control are lower than those on cars with average pollution control.

Sweden's current transport and fuel policies rely heavily on fiscal measures, including a modest carbon tax and fuel taxes differentiated on the basis of environmental quality. At least until the late 1980s, however, taxation of road fuels and new cars was less than the average for Europe. They were boosted on 1 July 1997 by SEK 0.25 (€ 0.03) per litre, raising Sweden's prices, which were already among the highest gasoline prices in Europe. Fuel taxes now automatically increase each year to maintain the tax constant in real terms.

Proposals to introduce road pricing or to vary other costs have met political resistance. A kilometre tax on diesel vehicles in effect for decades was lifted in 1993. As compensation, the tax on purchase of new vehicles using diesel was increased. No road pricing was introduced in the 1998 financial law, although the principle that each vehicle user should bear his or her full marginal social costs has been proposed as one basis for transport policy.

Sweden has supported the personal mobility of its citizens in ways that probably increase CO_2 emissions. By permitting tax deductions for commuting (and for using a car to commute where this saves a certain amount of time), Sweden makes it possible to change jobs without moving. Under a company-car scheme, many employees received car privileges in lieu of higher salaries, which was extremely attractive for tax purposes. By the mid-1990s, however, Sweden began to address the parts of company-car policies that allowed many drivers to escape the true impact of marginal fuel costs (and marginal driving).

Sweden recognises the potential importance of administrative measures, particularly those related to land-use planning and information. But these policy elements are playing a minor role in future CO_2 reductions, except at the margin. This may be because the three large cities (Stockholm, Gothenburg, and Malmoe), together with their suburbs, already have well developed bus and rail systems. Many analysts have given Sweden high marks for developing suburbs with rational land-use planning and transit availability. Development around transit nodes has been encouraged for decades. It may be that Swedes use public transit well and use their cars well, too.

Sweden has had an open-road policy for large trucks. Load factors and utilisation for trucks appear above average for Europe. Sweden (along with Finland) has the largest permitted truck gross weights in Europe, 60 tonnes. The government is now trying to ease the strain of trucking around cities through such measures as differentiated taxes and higher diesel taxes. But KOMKOM feels it is not feasible to increase taxes on large trucks to a level that fully reflects their external costs because of the impact this would have on freight costs in the near term. Moreover, high fees on Swedish trucking would make it more difficult for that industry to compete against foreign trucking. But the 1998 financial law makes it clear that transport policy will follow the decisions made by the European Union in this area.

While Sweden has long had broad transport policies, interest in fuel (and thus indirectly in CO_2) started with the first oil shock, in 1973-74. At that time, Sweden had the highest oil imports per capita of any country. A voluntary agreement took effect in 1978 aimed at reducing sales-weighted, new-car fuel intensity from more than 9.5 litres per 100 kilometres to 8.3. This goal was largely met at least as far as tests (rather than actual road use) are concerned. The 1979 oil crisis led to more measures to save oil: replacement of oil by electricity which is generated almost entirely by hydro or nuclear and renewables; increased efficiency in oil consumption for heating and industry; and reduced fuel use. All these measures resulted in the dramatic reductions in carbon emissions shown in Figure 6.1. Subsequent energy-policy measures dealt primarily with electricity and the nuclear-power controversy. The interest in saving road fuels seems to have subsided.

The 1988 Parliamentary Transport Policy was aimed at accessibility, efficiency, safety, a better environment (particularly in built-up areas), and regional balance. Between 1988 and 1991, a series of small but noticeable reforms in fuel taxation began to take aim at various pollutants in fuels. One such tax was converted to a CO_2 tax. After the 1992 Rio Conference, the Traffic and Climate Committee (TOK) focused on CO_2, while KOMKOM focused on overall transport policy reform, including changes in taxation. In KOMKOM's preliminary report, *New Course in Transport Policy,* a key goal was that each branch of transport should bear its marginal social costs (KOMKOM 1996). Estimates given suggested that cars only bear 80 percent of their fuel costs to society, trucks and busses 50 percent, and rail 20 percent. KOMKOM proposed no new taxes on biofuels for the present, but recommended that yearly taxes differentiated by weight be imposed on all vehicles, and environmentally-differentiated taxes on new vehicles (and on fuels). The KOMKOM proposal also called for maintaining a higher new-car tax on diesel vehicles because of diesel's higher external costs.

Environmental concerns about road fuels never subsided, however, as Sweden pushed for unleaded fuel and catalytic converters, and advances in other air-quality measures. The first CO_2-specific policy arose in the early eighties from the environmental tax reform (MIA) that transformed part of the tax on fuel into a CO_2 tax. But the first whole-hearted effort at confronting CO_2 came in 1994 when TOK issued the *Traffic and Carbon Dioxide* report (TOK 1994).

TOK focused principally on what fuel prices might meet CO_2 goals. It also examined other measures, such as CO_2 taxes, less fuel-intensive vehicles, taxation of company-car privileges, planning at the local level, regulation of sea and air fuel and a strong emphasis on research and development in alternative fuels. This was followed by KOMKOM, which examined the larger context of transport reform and tax reform within the transport sector. KOMKOM set a goal of 10 percent reduction in CO_2 emissions by 2010 and a 20 percent reduction by 2020, compared with 1990. It noted that 1995 was already 10 percent *higher* than 1990. For the longer term, it set a goal of a 60 percent reduction, but no year was given. How likely is this goal to be met? The 1998 Transportation Bill passed by Parliament aims at a target attaining the 1990 emission level again by 2010. Recent studies by the Swedish Institute for Transport and Communications Analysis (SIKA), a body in the Ministry of Industry, also regards this as a target.

In 1998, the Traffic Taxation Commission (Trafikbestkattningsutredningen, TBU), backed away from aggressive action. It agreed with the EU decision that environmentally differentiated taxation of cars was inappropriate. It suggested a flat tax on all cars, and proposed raising the tax on motorcycles, the tax on diesel fuel and the scrappage fee on older cars.

The Swedish government completed a transport policy bill in the spring of 1998, which was adopted by the Parliament (Riksdag) in June of the same year. The legislation embodied some KOMKOM proposals, but put off or rejected others. There were few concrete measures aimed directly at CO_2 in the bill. The overarching goal was to ensure long-term supply of transport for individuals and firms that would be socially and economically efficient. Five sub-goals are good access to transport, high-quality transport, improved safety, a good environment and a positive impact on regional development. Rather than setting its own standards, the government prefers to work through the European Union for lower fuel intensity, new environmental and safety classifications of cars, stricter exhaust and noise rules, and durability requirements for emissions controls.

Can more be achieved? KOMKOM's calculations assume that by 2010 about 40 percent more car-kilometres will be driven than in 1990. If there is to be a reduction in emissions of 15 percent, then the increase in kilometres driven requires a 40 percent decline in CO_2 emissions per kilometre. This decline implies a real fuel economy of roughly 6 litres per 100 kilometres or a large increase in the use of truly low-carbon fuels. In its base case, KOMKOM assumes a drop of 11 percent in fuel consumed per kilometre, from today's 10.1 litres per 100 kilometres to 9 litres per 100 kilometres. Examining these calculations, SIKA (1999) and Edwards (1999) find that, in addition to the European voluntary agreement, further increases in fuel prices should occur, from about SEK 8 (€ 0.93) per litre in 1999 to more than SEK 11 (€ 1.28) per litre in real terms. Edwards assumes only a 25 percent increase in total kilometres compared to 1990 and a 30 percent decline in the fuel intensity of the stock by 2010. This implies a much greater decline in the fuel intensity of new cars. The diesel share of new cars is estimated to rise to 24 percent by 2010.

For goods transport, the assumption is an increase of 30 percent in highway tonne-kilometres by 2020 over 1990. Given that the base case assumes a constant load of 13.1 tonnes per vehicle, this implies a similar increase in vehicle-kilometres. Here it is conceivable that an increase in loading (or decrease in empty back-hauls) could yield large reductions in vehicle-kilometres relative to tonne-kilometre, which, together with reduced fuel intensity in new trucks and some changes in tonne-kilometres might actually contribute to a large decline in emissions. The more recent modelling by Edwards assumes very small improvements in fuel intensity for trucking. SIKA suggests that improvements in load factors will make significant contributions to reduced emissions.

Selected Transport-CO_2 Initiatives

The coming years in Sweden will bring a modest package of fiscal stimuli, the tightening of company-car benefits, the Volvo voluntary agreement (which Volvo believes will be met without resource to biofuels), the application of the EU voluntary agreement, the moves to improve truck freight, an already rejuvenated inter-city bus system and a revitalised railroad. Together, they are likely to reduce the growth in CO_2 emissions. Growth in fuel use and emissions would no longer automatically track increase in GDP. Whether a net decrease occurs is uncertain, mostly because of the time it will take for the entire car fleet to be replaced.

Differentiated Taxes

Sweden has pioneered environmental classification for both fuels and vehicles. MK3 (environmental class 3) contains vehicles satisfying only the minimum requirements on fuel efficiency or on emissions. MK2 used to include vehicles or fuels that exceed these minimum requirements by a certain amount, as well as vehicles with long-lived pollution control equipment. In part because of Sweden's entry into the European Union, MK2 no longer exists in practice, and MK3 has become MK2, meeting an EU standard. MK1 includes only equipment of very high durability or "low emission vehicles," or "clean" fuels (and therefore particularly suited to built up areas). These requirements are satisfied by a few models (including some Volvos). After 1996, MK3 remained only in place for heavy trucks and busses, and these face increased purchase taxes. Between 1992 and 1996, nearly 60 percent of new vehicles belonged to MK 1 or 2, versus only 11 percent in 1992. The environmental taxes levied on the different classes of fuels and vehicles are shown in Tables 6.2a and 6.2b.

In 1997, the Traffic Taxation Commission concluded that EU rules on the use of environmentally differentiated taxes should be revised. They also proposed eliminating the longstanding yearly tax on light-duty vehicles

Table 6.2a. Fuel Environmental Taxes, January 1999 (in Swedish Krona, converted to Euros)

Fuel classification	Gasoline 2	Gasoline 3	Diesel 1	Diesel 2	Diesel 3
Energy Tax per litre	SEK 3.58 (€ 0.42)	SEK 3.65 (€ 0.43)	SEK 1.60 (€ 0.18)	SEK 1.824 (€ 0.21)	SEK 2.119 (€ 0.25)
CO_2 Tax per litre	SEK 0.85 (€ 0.10)	SEK 0.85 (€ 0.10)	SEK 1.049 (€ 0.12)	SEK 1.049 (€ 0.12)	SEK 1.049 (€ 0.12)

Source: ACEA 1999

Table 6.2b. Environmental Taxes on Passenger cars, excluding VAT, January 1999 (in Swedish Krona, converted to Euros)

Fuel	Gasoline	Diesel
Environmental class 1	No road tax during the first five years	No road tax during the first five years
Environmental class 2 & 3		
Up to 900 kg service weight[a]	SEK 585 (€ 68.50)	SEK 2223 SEK (€ 260)
901 kg service weight[a] *and above*	SEK 734 (€ 86) + SEK 149 (€ 17) per 100 kg above 1,000 kg	SEK 2790 (€ 326) + SEK 566 (€ 66) per 100 kg above 1,000 kg

a: service weight = curb weight + 70-kilogram driver
Source: ACEA 1999

based on weight. The commission found no environmental reason to justify a weight-based tax. It suggested a single tax of SEK 1,100 (€ 130) per year. It proposed that the yearly tax on diesel cars be raised to SEK 4,200 (€ 492) per year, a high increase for older cars.

For heavy vehicles (trucks and busses), the annual road tax depends on the gross vehicle weight, number of axles and the fuel used. From 1998, Sweden has joined the so-called Eurovignette system for heavy vehicles. This means a lower annual road tax for vehicles paying the Eurovignette fee.

Policies on Car Use for Business

The company car has played an important part of the rise in Swedish mobility, with as many as half of new cars purchased by companies for employee use. A widespread debate on the environmental impact of this policy has led to a tightening of rules. Legislation that took effect on 1 January 1997 limited the private benefits of company cars by increasing the beneficiaries' taxes. But there is still no direct tax paid on the marginal kilometres driven. The National Road Administration's own travel policy, which provides de facto guidelines to other government agencies, now states that company or private cars used for work may be reimbursed only if they use less than 8.6 litres per 100 kilometres in 1998 and 8 litres per 100 kilometres in 2000. This may rule out from eligibility some of the largest cars on the market. Drivers of such cars can only receive compensation after seeking an exception from the rule. (Westöö 2000).

Conclusions

In Sweden, past transport policy gives a mixed message: Big cars and open roads, and the highest per-capita auto use in Europe make CO_2 reduction difficult to achieve, yet the situation leaves much flexibility. The high fuel intensity of the Swedish car fleet – the highest in Europe – means that there is room for improvements in fuel economy, and that fact seems to have caught Volvo's attention. While Sweden's size and widely-spread population make for high per-capita mobility, this does not mean that technology cannot reduce emissions per vehicle or per passenger-kilometre.

Still, the policy debate has been difficult. The controversy over tightening company-car rules has led to several revised proposals from the government. Open roads – Sweden has less congestion than the continental countries studied for this report – also make it hard to push for forceful traffic reforms outside the large cities. Indeed, the recent difficulties encountered by Stockholm in solving its own transport problems suggest some rough sailing ahead. As for actual measures passed through early 2000, the revisions to company-car rules are the most outstanding. The Volvo VA has been followed by the more stringent EU-wide agreement. The principe that each transport mode should bear its full social costs has been accepted, but it remains to be seen exactly how each cost will be handled.

It is clear that the longer-term goals for CO_2 reduction from cars depend on the ability of the EU-wide agreement to influence the behaviour of companies, technologies, and car buyers. KOMKOM and TOK seem to aim for new-car intensities of about 6.3 litres per 100 kilometres, values well below those foreseen in the EU agreement. Biofuels may have to be added, which would mean that the CO_2 goal would be reached at a somewhat higher fuel intensity, but with lower CO_2-content fuels. Significantly, the 1998 law contains no specific quantitative goals for either fuel consumption or emissions per kilometre.

Sweden has begun to face a dilemma that is typical for a country with a large automobile industry and wide open spaces. While there is congestion in the largest cities, overall congestion, noise and air pollution is much less than in large cities of continental Europe. Under these circumstances, the average person may not feel threatened by transportation problems, and this can make it difficult to promulgate stringent policies to affect either transport or CO_2. KOMKOM shows that much more can be done. Technical measures and biofuels will both be crucial for reducing emissions.

The key elements for Sweden are a reduction in fuel use per kilometre for cars, which can be achieved through technical measures; restructuring of some transportation activity (and eventual restraint in the growth of vehicle use) which will follow from changes in taxation; and some "soft measures" that will restrain vehicle activity by an additional small amount. If biofuels prove successful and economic, their wide introduction on the market could make an important contribution. Because Swedish travel and freight volumes are high relative to GDP, it is conceivable that growth in basic demand for transportation will be only modest in Sweden, so that the measures mentioned here could have a real impact.

CHAPTER 7: THE UNITED KINGDOM

Background

Trends in Transport and CO_2 Emissions

In 1990, transport accounted for 24 percent of CO_2 emissions in the United Kingdom, a share that is expected to rise to 26 percent in 2000. According to the Department of the Environment, Transport and the Regions (DETR 1999a), total carbon emissions from the transport sector in 1996 were 34.6 megatonnes of carbon, of which almost 70 percent was from passenger travel. This is the equivalent of 0.57 tonne of carbon per person per year. Among the wealthiest nine countries of northwest Europe, this amount of transport carbon per person ranks fifth. Over the 25-year period between 1970 and 1995, passenger transport emissions grew by about 2.4 percent per year, compared with an annual growth of 1.5 percent for freight emissions. In 1995, more than 90 percent of CO_2 emissions in the passenger transport sector was from passenger cars, and 88 percent was from trucks in the freight sector.

In the United Kingdom, as elsewhere in Europe, a significant portion of transport sector emissions is associated with company cars. Official government statistics suggest that as much as 20 percent of vehicle-kilometres travelled is done in company cars, which represent more than half of all newly registered cars. To understand the impact of company cars on CO_2 emissions, it is important to understand what portion of that 20 percent of vehicle-kilometres travelled would have been avoided in the absence of company cars, or what portion would have occurred using other, potentially more polluting, equipment. Furthermore, even if company cars make up less than 20 percent of the fleet at any one time, former company cars make up a much higher share of the fleet, since

Table 7.1. Key Transport Statistics for the United Kingdom, 1997

Population (million)	**59**		
GDP per capita (US$ in PPP)	**21,170**	Cars (per 1,000 people)	**399**
Passenger transport activity (billion passenger-kilometres)	**721.1**	**Freight transport activity** (billion tonne-kilometres)	**180.9**
Passenger car	*88%*	*Road*	*85%*
Powered two-wheeler	*1%*	*Rail*	*9%*
Bus	*6%*	*Inland navigation*	*0%*
Urban rail	*1%*	*Oil pipeline*	*6%*
Railway	*4%*		
Transport final energy consumption (million tonnes of oil-equivalent)		**Transport CO_2 emissions** (millions tonnes)	
1990	**39.4**	1990	**114.5**
1997	**41.7**	1997	**121.2**
Rail	*3%*	*Rail*	*1%*
Road	*94%*	*Road*	*96%*
Inland navigation	*3%*	*Inland navigation*	*3%*
Pipeline	*0%*		

Source: *EU Transport in figures*, OECD, IEA

Figure 7.1. Trends in CO_2 Emissions in the United Kingdom, 1970-1997

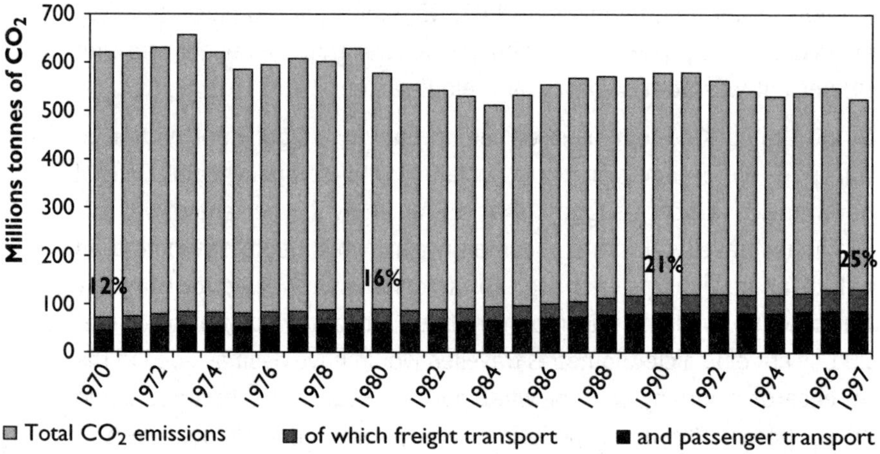

X% : Share of transport emissions in the total

Source: IEA/LBNL

company-car sales in the mid-1980s hovered around 50 percent of new cars. Industry spokesmen maintain that company cars are newer and company-car policies stimulate stock turnover towards "cleaner" vehicles with regard to hydrocarbons and nitrogen oxides. Yet the fact that company cars are newer (and therefore cleaner) has little long-term impact because eventually all cars are cleaned up, while this accelerated turnover only serves to put more cars into the used car market, thus increasing the overall size of the car fleet, and not simply forcing older cars out of the stock. Finally, many recent company cars have been diesel-powered, emissions from which aggravate some local pollution problems.

Figure 7.1 shows trends in CO_2 emissions in the United Kingdom from 1970 to 1997.

Public Attitudes and Perceptions

The British public seems to have an ambivalent attitude towards the car. While auto use is popular for most people in the United Kingdom, the British nevertheless recognise the destructive effect high levels of motorisation would have on an island the size of Great Britain. Consequently, although car ownership and use of the car has not slowed, the public by and large has thus far not been hostile to the previous and present governments' attempts to restrain car use. Whether and when a breaking point will be reached is an open question. Public opinion surveys around the time of the release of the transport White Paper in July 1998 showed that many were sceptical of the specific measures the government intended to introduce, yet there was general support for the aims of reducing car traffic. Most people in Britain were aware of the harmful environmental and social effects associated with transport.

As a measure of the general level of consensus in Britain, even groups that make up the traditional "road" lobby, such as the Royal Automobile Club or the Automobile Association, say they favour reducing car traffic and increasing expenditures on alternative modes. These groups are also sceptical of the details, in particular efforts to use fees on cars to cross-subsidise other modes, but they do acknowledge the need to discourage car use in general, particularly for the morning commute and the "school run," which by some estimates accounts for 18 percent of morning traffic in the United Kingdom, particularly within the larger cities.

Role and Position of Industry

The UK car manufacturing industry may no longer be the entirely British industry it once was; yet the companies that are currently active nevertheless have a British perspective on transport. Industry representatives (from subsidiaries of US companies) reflected the concerns – particularly acute in Southeast England – that traffic congestion was already so intolerable that measures to restrain car use are both inevitable and desirable, even from the manufacturers' point of view.

Although the programme itself is new, industry representatives were also quite positive about the Clean Vehicles Task Force (CVTF), which has been in place since March 1998. The CVTF not only provides a general forum for discussion of local and global pollution issues between industry and the government, but also serves as a significant source of information exchange during the government's ongoing major revision of the vehicle excise duty (DETR 1999b). In short, the car manufacturing industry is relatively positive in its outlook and contribution to the ongoing evolution of UK transport policy.

The same is not completely true of the oil and gasoline industry, which has quietly opposed the government fuel-duty increases. This industry does not dispute the need to reduce congestion, improve air quality in cities, and reduce overall carbon dioxide emissions, but it questions the government's motives and commitment in choosing to do so primarily through a tax on motor fuels. But despite such opposition, the industry does not appear to be actively working to reverse this tax policy, which is set to expire in 2002.

Other Influences and Trends

Because the United Kingdom is a country situated on islands, separated from mainland Europe, it has a certain margin for manoeuvre independent of Brussels – a margin many of its European neighbours do not have. It has no cross-border freight transit traffic, and no direct cross-border car traffic. This means that unilaterally raising fuel taxes is less problematic in the United Kingdom than elsewhere in Europe, and the British are certainly using this geographic distinction to advantage.

Evolution of Transport-CO_2 Policy

Like many countries in the industrialised world, the United Kingdom following World War II pursued a transport policy that emphasised motorised transportation – in particular, the private car. This policy orientation was strengthened in the 1980s. Despite renewed focus on reducing public sector spending, public investments in the road network – particularly un-tolled motorways funded through general revenues – grew during this decade. At the same time, overall public spending for transportation was reduced, and investments in alternatives came to a halt. The groundwork was laid during this period for privatising much of the "public" transport network; most of Britain's urban bus networks and all of its inter-city rail network have since been privatised.

Recently, the consensus of what constitutes good transport policy, as well as a sober inventory of what is or is not possible with current and likely future budget constraints, changed, making the current stated policies markedly different from those just described. There now appears to be an understanding that unrestrained motorisation and car use is not economically, socially, or environmentally sustainable on an island as small as Britain. As early as 1993, the government had put forward a policy of a 3 percent increase per year in the tax on motor fuels – partially as a revenue measure, and partially in response to the Rio agreement, which was negotiated the previous year. A Green Paper issued in 1996 stated that growth of car traffic was unsustainable and that efforts to restrain car growth would be necessary (DETR 1996). This Green Paper reflected many policies that were already being observed, including the annual fuel-duty increase, which the government by this time had increased to 5 percent per year over inflation. The Green Paper not only marked a significant change in *de jure* policies but also set the stage for more far-reaching efforts by the current government – changes in pricing, changes in the degree of scrutiny for roadway projects, and new emphasis on other transport modes besides the car – as outlined in its July 1998 White Paper.

Transport policy in the United Kingdom began to shift in other important ways also. Motorised transport was favoured by the government for many years, together with expenditures dedicated to the development of roads. The 1994 Standing Advisory Committee on Trunk Road Assessment (SACTRA) report, however, concluded that trunk roads induce traffic and can lead to an increase in traffic (SACTRA 1996). This, combined with the enormous cost of the programme, caused the highways programme to be progressively scaled down on three different reviews, and led to significant changes in practice.

The second initiative – reducing car dependence – included three important measures. First, the government issued important planning and policy guidance on integrating transport and land-use planning to reduce car dependence, as well as strengthening the traditional centre city core. Second, it put a lot of support behind local initiatives known as "Travel Awareness" schemes – locally based public relations and awareness efforts to get people to reduce their car use (for example, TravelWise in Hertfordshire or Headstart in Hampshire). In the United Kingdom (particularly in England and Wales), however, land-use plans are not legally binding on districts and counties, so while there are often political pressures for local legislators to adhere to them, strong economic or development pressure may cause district or county councils to ignore the plans entirely in issuing building permits. Still, local authorities increasingly are linking planning permission to the production of "green" transport plans. In addition to these more general planning efforts to reduce car dependence, the Green Paper also acknowledged for the first time that traffic management and restraint measures would be needed, leading to the third important measure of support for reducing car dependence: the Road Traffic Reduction Act of 1997. Many of these ideas were codified into the 1998 White Paper on integrated transport, *A New Deal for Transport*.

The United Kingdom has one of the most comprehensive environmental and transport strategies in Europe. Its multi-faceted approach is a combination of several important elements. Perhaps most significant of these was the commitment to raise taxes on automotive fuels by, on average, 5 percent per year above the rate of inflation – a commitment that was increased in March 1998 to 6 percent per year. Since 1993, gasoline taxes have increased 6.75 percent per year in real terms, and diesel taxes have increased 8.25 percent per year. These increases, however, have not translated directly into proportionally higher fuel costs at the pump of the same magnitude, because of the strength of the pound sterling and because world petroleum prices continued to fall during the period. The government had estimated this initiative would save 3 megatonnes of carbon by the year 2000, and 7 megatonnes by 2020.

The Government has developed a number of policy instruments under the heading of "Greener motoring."

In 1997, the government published a *Guide to Green Transport Plans*, reinforced with *Changing Journeys to Work*. This measure recommends that companies reduce by 10 percent the total number of people commuting alone to and from work by car. This programme would be first implemented in all government departments and would cover not only commuting but also travel in the course of work (fleet management).

Use of greener vehicles is promoted by the Clean Vehicles Task Force. This takes the form of a partnership between government and industry to look at ways of greening vehicles both by encouraging the production and purchase of more fuel-efficient, less polluting new vehicles, as well as encouraging measures to cut emissions for the existing fleet. The measures to support greener transport constitute a major change in the tax system.

The Road Traffic Reduction Act (RTRA) of 1997 directs local county authorities to estimate existing and anticipated future levels of traffic, designate a reduction target, and enumerate measures to reach the target. Unlike the US Congestion Management Plan (part of the Clean Air Act Amendments of 1990), which requires only non-attainment areas to reduce traffic, the RTRA requires all localities to do so, though it gives them leeway in establishing the reasons for the traffic reduction, and thus, in selecting measures for implementation. Most counties will probably target congestion, air quality, or protection of particular sites from excess road traffic in considering their RTRA targets and strategies. However, if successful, these strategies will also reduce CO_2, because the act addresses motor vehicle activity itself, instead of individual impacts such as noise, congestion, and air quality.

When implemented, the initiatives included in *A New Deal for Transport* might also have a significant impact on CO_2 emissions. This White Paper sets a framework within which detailed policies will be taken forward. Some of the proposals will require legislation, and there remains some question as to how much of the programme will actually be implemented. Some critics from environmental advocacy groups, for example, point out that many of the more important, even revolutionary, initiatives, such as congestion or road pricing, are left up to local governments. Others question how significant a role public transport can play. Public transport accounts for only about 14 percent of all trips, and this number has been in decline for several decades.

Selected Transport-CO_2 Initiatives

Increase in Fuel Taxes by 6 Percent per year over Inflation

The fuel duty strategy has been seen as the main tool to reduce emissions from road transport. Since 1993, fuel taxes in the United Kingdom have risen faster than inflation. These increases occurred with little apparent opposition from the public, at least until 1999, when public pressure mounted to moderate future increases.

Higher fuel taxation can be an effective tool to encourage drivers to reduce consumption both by reducing their mileage and by using more efficient vehicles. According to a DETR consultation paper published in October 1998, the impact of the fuel tax increases between 1996 and 2002 is estimated to save from 2 to 5 megatonnes of carbon in 2010

(DETR 1998b). DETR notes that new-car fuel economy has improved in recent years, which they attribute in part to the somewhat higher prices. They point out that the knowledge that fuel taxes would continue to increase in the future could have played an important role in consumers' decisions on which cars to buy. However, other views differ on the impact of the tax measures. According to the oil industry, consumer responsiveness to fuel prices is too low in the short-to-medium term for the tax increases to have had much of an effect on fuel demand.

Given the low elasticity of demand for motorised travel, a small one-time increase in fuel taxes may not have much of an influence on the behaviour of drivers. On the other hand, as a long-term strategy, gradual and steady increases in fuel taxes make sense as part of a broader package of measures, particularly when fuel prices are falling. The UK has attempted to use the fuel price "lever" in just this manner – as the centrepiece of a broader programme to discourage auto use and encourage other modes of travel.

Variabilisation of Charges Associated with Car Use

The White Paper *A New Deal for Transport* discussed a number of initiatives to charge for car use – that is, increasing variable costs both absolutely and relative to fixed costs. Among the various charging schemes under consideration were the following:

- Road pricing, congestion pricing, and cordon pricing for city centres.
- Employer-provided parking taxed as income.
- Taxation on parking provided at out-of-town shopping centres.
- Toll charging on motorways.

There had been some anticipation before the release of the White Paper that these initiatives would be adopted at the national level. Instead, the British government's policy simply indicated that it would make these charges available to local governments that choose to incorporate them into their local transport plans as part of a package of measures.

In the end, the latter two charging proposals – taxing of out-of-town shopping parking and toll charging on motorways – were eliminated from the White Paper, but the government left open the possibility of introducing motorway charges through further consultation. The other two proposals were devolved to local initiative. Why these changes from what had been expected occurred has been the subject of speculation. The fact that two important measures have been dropped and that the local governments are taking responsibility for the initiative is likely to make the White Paper less potent than at first expected. Threat of competition between jurisdictions will make local jurisdictions wary about using these expanded powers. Nevertheless, that local governments will be allowed to implement road-pricing initiatives and to tax employer-provided parking is an important step in shifting fixed costs of running a car to variable costs. How widespread road pricing will become remains to be seen; at present, only some of the more progressive local governments seem to be actively considering it. But the government has not ruled out the possibility of national government-sponsored road pricing in the future.

Green Motoring and the Clean Vehicles Task Force

The Clean Vehicles Task Force (CVTF) was formed in 1998 to address the following:

- Strengthening compliance and enforcement of exhaust emission standards.
- Improving the fuel efficiency of new vehicles.
- Promoting purchases of greener vehicles.
- Developing advice for local authorities wishing to establish low-emission zones.

The CVTF is still relatively new, but both representatives of government and industry have expressed satisfaction with the tenor of the meetings held so far. It seems to be more productive and less rancorous than previous efforts or, for example, the "Car Talk" initiative in the United States.

The Task Force published its first report in July 1999. The report recommends action to inform consumers, improve enforcement, support fleet operators and promote technological solutions through the adoption of measures presented in Table 7.2.

In March 1998 the government announced changes to the Vehicle Excise Duty (VED), to take into account the relative environmental impacts of different kinds of vehicles. The new system with graduated taxation was put in place on 1 June 1999, lowering the duty to £100 for cars with a maximum engine size of 1,100 cc, with the taxation level remaining at £155 for other vehicles. From March 2001, the rate will be increased to

Table 7.2. Clean Vehicle Task Force Recommendations

Inform consumers by:	Providing better information on fuel consumption*, emissions and noise. Developing a clear vehicle label showing environmental information. Promoting improved maintenance and better driving styles. Encouraging regular emission testing at minimal cost in standard vehicle servicing. Increasing low-cost emissions testing facilities, and developing self-testing for emissions. Developing effective on-board driver information systems to give data on emissions.
Improve enforcement by:	Developing road-side emission testing to target the worst polluters effectively. Improving the MOT** emissions test. Developing low emission zones, to improve air quality in urban areas.
Promote technological solutions by:	Encouraging retrofitting for existing vehicles. Promoting alternative fuels and the infrastructure to supply them. Supporting research and development into alternative fuel sources, new power sources, and other technologies.
Support fleet operators by:	Developing a greener fleet certification scheme. Encouraging the adoption of voluntary targets. Providing best practice guidance.

Source: DETR 1999b
* See DETR 2000a
** MOT is the annual roadworthiness test in the United Kingdom

£105 and extended to cars with an engine size below 1,200 cc, other vehicles being taxed £160. The government recently announced that from March 2001 new cars will be taxed according to CO_2 emissions (DETR 2000a). The new system of VED, based on fuel and CO_2 emissions (see Tables 7.3 and 7.4), will send a better signal to vehicle manufacturers and purchasers about the environmental impact of the cars they make and use, and will encourage the use of more fuel-efficient cars. CO_2 is a priority, but an efficiency strategy will also provide some benefits in terms of reducing local pollutant emissions.

Some car industry representatives have commented that this initiative represents an important symbolic gesture – and were thus rather supportive of it – although they felt that it would have little impact in real terms, because the VED represents such a low percentage of overall lifetime ownership and operating costs of vehicles.

Table 7.3. Vehicle Excise Duty Applicable to New Cars Registered on or after 1 March 2001

Band	CO_2 Emissions (g/km)	Alternative Fuel Car	Gasoline Car	Diesel Car
A	Up to 150	£90	£100	£110
B	151 to 165	£110	£120	£130
C	166 to 185	£130	£140	£150
D	Over 185	£150	£155	£160

Source: http://www.dvla.gov.uk/gved/gved.htm

Table 7.4. Example of Vehicle Excise Duty Paid for a Ford Focus 5-Door Saloon

Motorisation	Fuel	Band	VED
1.8 Tci	Diesel	A	£110
1.4i 16V	Gasoline	B	£120
1.8i 16V	Gasoline	C	£140
2.0i 16V	Gasoline	D	£155

Source: http://www.dvla.gov.uk/gved/gved.htm
Note: The choice of the car model and make is meant as an illustration and is based on the Driver & Vehicle Licensing Agency Information, as available in May 2000.

Conclusions

The British are focusing their transport policy on adjusting price signals in combination with supporting measures such as promoting "best-practices." Road pricing has been left to the decisions of local, inter-jurisdictional competition, which means that, even where implemented, its true effects will be hard to gauge. Measures to encourage environmental purchases via changes in the VED or reform of taxation of company cars are more likely to be comparable to environmental labelling (still potentially important), rather than fiscal measures, because the costs of such incentives relative to lifetime costs are so low. Finally, the fate of the various initiatives and ideas contained in the White Paper remains uncertain.

DETR itself strongly supports the idea that one of the most significant measures for reducing CO_2 emissions is the voluntary agreement recently concluded between ACEA and the European Union, whereby the former agreed to a fleet-wide average standard of 140 grams of carbon per kilometre by 2008. DETR's traffic forecasts project significant reductions in carbon emissions resulting from this agreement, with the annual increases in fuel duties, (especially if they are continued after 2002), representing a complementary policy that helps ensure that the decreases in carbon intensity guaranteed by the ACEA voluntary agreement get fully translated into reductions in carbon emissions, with little rebound effect due to lower fuel costs as fuel efficiency improves.

While there were significant gaps between what was under discussion and what was finally included in the transport White Paper, taken as a whole, the array of policy options at the government's disposition for both transport policy in general, and CO_2 reduction in particular, is far reaching and consistent. For this reason, the United Kingdom is a country worth observing for the next several years. In particular, it will be interesting to see whether the various policy initiatives – the annual fuel tax increase, the changes in the vehicle excise duty, and the initiatives announced in *A New Deal for Transport* – will differentiate the United Kingdom from other countries participating in the EU/ACEA voluntary agreement.

CHAPTER 8: EUROPEAN UNION

Background

Trends in Transport and CO_2 Emissions

Collectively, the 15 Member States of the European Union (EU) have 40 percent more people than the United States and a GDP that is almost equivalent. Per capita vehicle ownership is still much lower than in the United States, as is the average distance travelled by an average EU citizen (around 12,000 kilometres per year versus 23,000).

Freight transport in the EU is much less GDP intensive than in the United States. In the European Union, the share of transport CO_2 emissions relative to total CO_2 emissions increased from 20 percent in 1990 to 26 percent in 1996 (EC 2000b). The reference emission level in 1990 was 733.8 megatonnes of CO_2 (EC 1997a). Analysis shows that in the absence of new policy measures, it is the sector with the greatest potential growth in CO_2 emissions up to 2010. As a reference, emissions of CO_2 were 825.4 megatonnes in 1996, which is equivalent to an annual growth of 1.9 percent over the 1990-1996 period.

Not surprisingly, road transport accounts for the major part of the transport emissions. In 1997, cars accounted for about 50 percent of transport CO_2 and road freight for about 35 percent. Half of all road transport emissions are the result of traffic in urban areas. From 1985 to 1996, road transport CO_2 increased by nearly 38 percent. When accounted for, air traffic within the EU generates 12 percent of transport CO_2, but these emissions grew 65 percent over the same period, and are forecast to grow at 6 percent annually in the medium term. Rail, inland waterway, and maritime transport are much less energy-intensive, and their emissions are less important. So road transport and air transport are

Table 8.1. Key Transport Statistics for the European Union, 1997

Population (million)	373		
GDP per capita (US$ in PPP)	21,286	Cars (per 1,000 people)	454
Passenger transport activity (billion passenger-kilometres)	4,511 *	**Freight transport activity** (billion tonne-kilometres)	1,641 **
Passenger car	*82%*	*Road*	*73%*
Powered two-wheeler	*3%*	*Rail*	*14%*
Bus	*8%*	*Inland navigation*	*7%*
Urban rail	*1%*	*Oil pipeline*	*5%*
Railway	*6%*		
Transport final energy consumption (million tonnes of oil-equivalent)		**Transport CO_2 emissions** (millions tonnes)	
1990	231.4	1990	655.8
1997	258.6	1997	723.7
Rail	*3%*	*Rail*	*1%*
Road	*94%*	*Road*	*96%*
Inland navigation	*3%*	*Inland navigation*	*3%*
Pipeline	*0.02%*		

* European air traffic (not included) was 322 billion passenger-kilometres in 1997.
** Sea transport intra EU (not included) accounted for 1124 billion tonne-kilometres in 1997.
Source: *EU Transport in figures*, OECD, IEA.

the two main focuses for reducing CO_2. In its pre-Kyoto scenario, based on past trends and policies not allowing for further reductions, the European Commission (EC) estimates that CO_2 emissions from transport (notably road and air transport) would continue to rise strongly, increasing about 40 percent between 1990 and 2010 (EC 1997a). Therefore, transport's tendency to continuous growth is a challenge for the achievement of any emission reduction target.

Under the Kyoto Protocol, the EC and the Member States have commitments to reduce GHG emissions by 8 percent from 1990 levels in the period 2008-2012, and have entered into a complicated set of national emission targets that distribute growth or reduction in carbon emissions among the 15 Member States. A key provision of the Protocol for the EU allows the EU and the Member States to fulfil their commitments jointly through a differentiated commitment between Member States (burden sharing).

Role of the European Commission

The European Commission has played a growing role in transport, environment, and CO_2 policy in Europe mainly due to the increasingly international dimension of transport. Trade patterns, business operations, and problems related to pollution and congestion occur to varying degrees without regard to national borders. The Kyoto agreement entered into by the European Union as a whole, and the transport emissions that are a major obstacle to achieving the targets, oblige the European institutions to take new, co-ordinated steps in the common transportation policy. These steps are needed to enable the Union and the Member States to achieve the Kyoto target together with, and as a complement to, national policies and measures.

Developing and implementing co-ordinated policies and measures is complicated by the distribution of authority between Member States and the Commission. Legislative action on an EU-wide level is only taken when measures at a national level are insufficient or inappropriate, or when harmonisation is required across the Union. There is a complex process of identifying, defining, and agreeing on measures between Member States and the European institutions when Union action is considered necessary and possible before specific actions are finally agreed upon. Furthermore, energy policy is not a responsibility of the European institutions. Where the question is addressed by the EC however, it is always through other policy areas, such as environmental protection. Thus, the Commission can take action only in a few policy areas that are outlined in its own policy strategy papers. In other areas, the Commission has to call on national, regional, and local actions to achieve the common target. Where EU-wide action is agreed upon, the Union's legislative framework then establishes (often minimal) standards and requirements, which are then enacted by legislation in the Member States.

Evolution of Transport-CO_2 Policy

Since issuing its White Paper on the future development of a common EU transport policy in 1992 (EC 1992), the European Commission has put forward a number of Green and White Papers that serve as a basis for discussion and as a vision for transport policy in the Union and the Member States: *The Future Development of the Transport* Policy (EC 1992), *Towards Fair and Efficient Pricing* (EC 1996c), *The Citizens' Network: Fulfilling the Potential of Public Passenger Transport in Europe* (EC 1996b), *A Strategy for Revitalising the Community's Railways* (EC 1996a), and the Directive "Charging of Heavy Goods Vehicles for the Use of Certain Infrastructures" (EC 1999b). These papers, as well as strategy papers for other sectors, such as *Energy for the Future: Renewable Sources on Energy* (EC 1997d), which establishes targets for biofuels, serve as an umbrella for discussion of common targets and areas where legislative action on the EU level is needed. With regard to CO_2 emission mitigation, three discussion papers from the Commission have tried to distil the Union's approach by combining policies already brought forward with new, genuine CO_2-targeted measures that are under development: *Climate Change: The EU Approach for Kyoto* (EC 1997a), *Climate Change: Towards an EU post-Kyoto Strategy* (EC 1998a), and *On Transport and CO_2: Developing a Community Approach* (EC 1998b).

The Commission's statements on transport and CO_2 (EC 1997a and EC 1998b) try to give a comprehensive picture of where emission reductions could be achieved in the future. They increasingly show that the integration of environmental objectives into sectoral policies has become a prime objective in European policies. Infrastructure-related measures and measures supporting modal shifts and intermodality of road and rail/ship are included, along with measures targeted at vehicles fuel efficiency, and behavioural/organisational aspects (Table 8.2). The long-term impacts of infrastructure provision (under the Trans-European Networks framework for Transport – TEN-T) and urban and land-use planning policies, are acknowledged and put into the context of CO_2 emissions. The integration of measures (pricing and taxation with regulations and demand-management measures) is a key concern.

Finding efficient ways to minimise market distortions caused by externalities appears to be central to the Commission' strategy, with considerable emphasis put on electronic road pricing in the future. These broad outlines are not representative of actual EU-wide legislation. The Commission's vision as defined in their latest communications (EC 1998b and EC 1999a) includes many measures that need to be taken on the national levels. The following elements constitute this framework:

Road freight. Best practices, such as improved logistics and more efficient freight operations, could achieve a reduction in truck operations and cut the number of kilometres travelled by 10 to 40 percent, with a significant decrease of CO_2 emissions. Increase in utilisation of vehicle capacity, reduction of empty running and driver training are solutions highlighted by the Commission.

Cars. Average fuel consumption and, therefore, emissions are increasing after a decline in the 1980s because of the trend towards heavier, more powerful cars. The solution lies in technical improvements and a shift towards smaller vehicles. The Commission's strategy consists of a voluntary agreement on fuel efficiency of passenger cars with the European car manufacturers, support of fiscal measures (such as differentiation of vehicle taxation) and a consumer information scheme to induce demand towards more low-consumption vehicles. The voluntary agreement with the European Automobile Manufacturers, which is a key action in the Community strategy to reduce CO_2 emissions, is discussed in further detail below.

Rail freight. Railways can potentially carry a larger share of freight. The revitalisation of this sector is a crucial dimension of European transport policy. Introducing market forces through deregulation and liberalisation is considered necessary to improve rail services (EC 1996a). Three policy packages are currently under discussion. The first contains guidelines for the use, management, and pricing of rail infrastructure; the second establishes rules of the financial relationship between the Member State and the railways; and the third sets out an approach to further technical harmonisation and interoperability in conventional rail. These measures would benefit greatly from a gradual extension of access rights in rail

Table 8.2. Common Transport Policy Measures for Cutting CO_2 Emissions

	Areas Affected	Policy Instruments	Actors
Infrastructures	• Land-use planning	Environmental impact assessments	Member States / Local Authorities
	• Infrastructure investments, urban transport, trans-shipment facilities, rail improvement, combined transport development	Structural funds, Trans-European Networks (TEN), Strategic Environmental Assessment	MS/LA + European Union
	• Infrastructure charging	Road taxes and different forms of road pricing	MS + EU
Fuel & Vehicles	• Progressive technical improvements of vehicles		
	- Exhaust and noise emissions, fuel consumption, performance, final disposal	- R&D - VA - Regulation - Vehicle testing - Recycling of parts - Fiscal incentives	- Industry + EU - Ind. + EU - EU + MS - MS + EU - Ind. - MS + EU
	• Composition & consumption of fuels		
	- Alternative fuels, cleaner fuels - Complete move to unleaded petrol by 2000	- R&D - VA - Fiscal incentives - Regulation	- Ind. + EU - Ind. + EU - MS + EU - EU + MS
Use Patterns	• Promotion of environmental use of private cars and trucks		
	- Driver information & education on a more rational use of car / trucks	- Campaign in media	- LA/MS +EU
	- Speed limits and other physical constraints	- Networks (car-free cities, cycling campaign)	- LA/MS +EU
	- Improved logistics	- Speed limits and other physical constraints	- LA/MS + EU
	- Development of car sharing	- Lower tolls, lane restrictions	- LA/MS
	• Improvement of the competition of environment-friendly modes		
	- Improved public/collective transport	- Investments, land-use planning - Rail liberalisation - Intermodality	- LA/MS + EU
	- Promotion of cycling and walking	- Campaigns in media	- LA/MS
	- Discouraging road traffic in cities, development of park-and-ride facilities	- Charges, high parking fees within cities	- LA/MS
	• Development of interactive communication infrastructures	- Logging and tracking systems - Teleworking - Video conferences	MS + Ind. + EU

MS: Member States; LA: Local Authorities; EU: European Union; Ind.: Industry
Source: EC 1992, 1994, 1998b

freight as outlined in the Commission's communication on the state of the railways.

Public Passenger Transport. Public transport use could do much to relieve urban congestion. By identifying innovative schemes already operating, the Green Paper *The Citizen's Network* (EC 1996b) pointed the way forward for urban transport. It states that the development of public passenger transport systems should be given greater priority if further adverse consequences for the citizen's quality of life and for the environment are to be avoided. A forthcoming paper is supposed to show how a modernised regulatory framework can be created at the Union level. National, regional, and local authorities and the private sector are called upon to review the effect of subsidies and other financial and fiscal advantages, such as free parking or company cars, on traffic congestion and the under-utilisation of public transport.

Air Transport. Although aviation's share of transport carbon emissions is still relatively small, the growth in aviation is two to three times higher than the average growth in transport. Policy options include stricter international emission standards, taxation and charging to improve the efficiency of the Air Transport Management system, and policies to develop alternatives to aviation, where appropriate. Taxation of jet fuel is proposed under the new proposal for a Community framework for taxation of energy products, though harmonisation in a wider international context is also considered necessary.

Fuel Taxation. In its proposal for setting up a Community-wide framework for the taxation of energy products (EC 1997b), the Commission aims at harmonising fuel taxation practices in Member States in two different ways. First, the minimum fuel taxation levels for transportation fuels that were introduced earlier have been considerably increased recently and extended to cover all energy products (including natural gas) used in transport (Table 8.3). Second, the currently requisite approval by the Commission of national exemptions (as for public transport companies) for fuel consumption will be generally allowed. National authorities will also be allowed to apply lower fuel duties to encourage the use of alternative fuels with a lower global warming potential than conventional fuels.

Table 8.3. Proposed Minimum Levels of Taxation Applicable to Motor Fuels

	1 January 1998	1 January 2000	1 January 2002
Petrol (ECU per 1,000 litres)	417	450	500
Gas Oil (ECU per 1,000 litres)	310	343	393
Kerosene (ECU per 1,000 litres)	310	343	393
LPG (ECU per 1,000 kilograms)	141	174	224
Natural gas (ECU per gigajoule)	2.9	3.5	4.5

Source: EC 1997b

For many EU countries that have national fuel duties well above the minimal levels, this does not mean a great change. Indirectly, it might give some countries more range of domestic manoeuvre in fuel taxation, if the difference in fuel duty to neighbouring countries, that creates significant fuel tourism, is reduced. Furthermore, more flexibility in differentiating taxes by fuels and users, as well as complementing or partly replacing fuel taxation with road pricing, is foreseen.

Fair and Efficient Pricing. In the White Paper *Toward Fair and Efficient Pricing* in Transport (EC 1996c), the Commission acknowledged that transport taxes and charges are currently set in many different ways across modes of transport and that they do not fully cover external costs in general. This can result in a distortion of competition between modes and create obstacles for the development of an integrated transport system, since it does not give users and manufacturers incentives to adjust their transport behaviour. Advanced technology for electronic road pricing, which is now becoming available is understood to be one of the key elements for "fair and efficient" pricing. It could partially replace fuel taxation, which is a relatively blunt, undifferentiated (in terms of traffic location and time-of-day) measure. To achieve "fair and efficient pricing," a directive was introduced in 1999 to harmonise vehicle taxes and infrastructures charges in line with the "user-pays" principle. Its main objective is to eliminate distortion in competition among haulage firms. First, the directive sets out, on a country-by-country basis, the vehicle charges envisaged (each Member State must levy and collect the taxes). Second, the Directive lists the conditions to be met by Member States wishing to introduce (or maintain) tolls and users charges (EC 1999b). Finally, the Directive suggests that in addition to these taxes, the Member States may apply "taxes or charges levied upon registration of the vehicles or imposed on vehicles or loads of abnormal weights and dimensions, parking fees and specific urban traffic charges and charges aimed at combating road congestion and the resulting pollution." The implementation of the legislation by each Member State will be completed by 1 July 2000.

Creation of a Community Transport System: Intermodality, Combined Transport and Logistics. In a systems approach, more effective use of the existing capacities will be encouraged. For this purpose, the Commission is developing an Action Programme (EC 1997c). This would encompass the increased use of information systems developed and tested in recent years (integrated logistics management systems, traffic management systems) and also of terminals to tranship goods. With the help of the "freeways," combined transport could significantly reduce the progression of long haul trucking.

Selected Transport-CO_2 Initiatives

The overview given in the Commission's strategy paper for transport and CO_2 (EC 1998b) summarises potential areas for action and estimates the potential for emission reduction. In many areas, however, no clear indication is given how these reductions can be achieved, either on a Union or on a national level (the recent communication [EC 2000a] on policies and measures to reduce greenhouse gas emissions gives a list of such actions by sector).

The Commission is in the difficult position that – having negotiated an overall reduction of 8 percent – it has to rely on each individual Members State to meet its target as agreed under the burden-sharing agreement. Clearly, non-compliance by one or more Member States is likely to have an impact on overall compliance, which is the Community's responsibility.

So defining the respective roles of the Community and the Member States is difficult. Furthermore, there is often a mismatch between what countries expect from the Commission (for example, as displayed in their national communications to UNFCCC) and actual support to a Union-wide decision on certain subjects. In this sense, it is important to look at the policy areas where the Commission has direct influence and where concrete legislative proposals are made for measures to abate emissions on a Union-wide level.

Transport Initiatives that Indirectly Affect CO_2 Emissions

European Union policy has an impact on transport, and thus indirectly on CO_2 emissions, through the following:

- The research and development programmes led by the Industry, Transport, Environment, Research, Telematics, and Energy Directorates.
- Market (de-)regulation measures, in particular for freight and rail.
- Co-ordination of planning and construction of road, rail, and information infrastructure within the Trans-European Network (TEN) plan.
- Harmonised (minimal) fuel taxation for vehicle fuels.
- Development of the concept of sustainable mobility and intermodality.

Voluntary Agreement on Fuel Efficiency

In 1998, the EU Council approved the voluntary agreement on fuel efficiency for passenger cars with the Association of European Car Manufacturers. This agreement, which may require a major move away from larger, heavier, higher-powered cars towards more fuel-efficient designs, is the only recently-negotiated measure with a CO_2 focus.

Under the ACEA umbrella, European car manufacturers have negotiated a voluntary agreement with the Commission that has helped to avoid implementation of mandatory fuel efficiency standards. The Commission's original target was 120 grams of CO_2 per kilometre for an average of new cars sold in the 2005-10 time period, and the negotiations in 1998 moved towards 140 grams of CO_2 per kilometre as the average emission level of cars sold in Europe by the year 2008 (EC 1998d). ACEA gives a target value for 2003 of 165-170 grams of CO_2 per kilometre (ACEA 1998) in order for the final target to be achieved. The value for 1995 was 186 grams of CO_2 per kilometre. In 2003, progress will be assessed and reconsidered in view of the possibility of achieving 120 grams of CO_2 per kilometre by 2012. Table 8.4 gives estimated levels of what the target in CO_2 terms represents in fuel efficiency of vehicles (assuming no fuel switching). Furthermore, ACEA had made a commitment that cars with less than 120 grams of CO_2 per kilometre would be available on the market by the year 2000.

ACEA's proposal engages the European car manufacturers as a whole – not individual manufacturers – to achieve the target. Technological improvements in vehicle efficiency, as well as "structural shifts" such as increased use of diesel and downsizing of vehicles (or more likely, "down-weighting" and perhaps "down-powering," or at least not "up-powering"), are supposed to play a role. No direct sanctioning mechanism is foreseen within the current proposal, but the EU has indicated that non-achievement could lead to the implementation of mandatory fuel-efficiency standards.

Table 8.4. Conversion of Target for Specific CO_2 Emissions of New Passenger Cars into Fuel Consumption

	Gasoline			Diesel		
g CO_2/km	l/100 km	km/l	miles per gallon	l/100 km	km/l	miles per gallon
180	8.0	12.5	29.6	6.7	14.9	35.2
160	7.1	14.1	33.3	6.0	16.7	39.6
140	6.2	16.1	38.1	5.2	19.1	45.3
120	5.3	18.8	44.4	4.5	22.3	52.8

Notes: Conversion factors according to IPCC Guidelines; assumed densities: 0.75 kg/litre gasoline, 0.835 kg/litre diesel.
The conversion is approximate; actual values can vary depending on specific fuel properties

The extent to which fuel switching, especially increasing use of diesel, might play a role is unclear. A key question is whether light-duty diesel engines will be able to meet the tough EU pollutant-emission regulation for 2005 (Euro-IV) with the gasoline and diesel fuels that will be available at that time. In fact, ACEA's commitment makes the availability of "enabling" fuels a precondition for success. Planned changes in future fuel specifications have been defined in the European Auto-Oil II process (EC 1996d), but it is unclear whether these will go far enough to be "enabling," especially in terms of sulphur content.

If diesels are able to meet the Euro-IV standards, their importance in meeting the agreement could be substantial, although the use of a CO_2 reduction target, rather than a fuel intensity target reduces the advantage of the diesel engine somewhat, since it produces more CO_2 per unit of energy consumed than does gasoline (but still less per kilometre).

Other options for increasing fuel efficiency, such as direct-injection, lean-burn technology in gasoline engines, can also hamper efforts to reduce pollutant emissions from motor vehicles. In particular, the NO_x emissions standards for passenger cars (Euro-IV in 2005), and the ability of manufacturers to employ very low NO_x catalysts (which require low-sulphur fuels), may have an impact on the choice of technologies to lower fuel consumption in gasoline vehicles. Thus, fuel specifications (especially sulphur and aromatics content) will be important for gasoline as well as diesel fuels.

The extent to which market shifts and the increased role of diesel play an important role could also depend on accompanying government actions to support them, such as the tax differential between competing fuels and between vehicles with different fuel efficiencies. In this sense, the agreement is not specific about the role of individual governments in assisting ACEA in meeting their targets. However, the EU framework specifies a "third leg" (to go along with the voluntary agreement and fuel economy labelling) to investigate other potential policies to encourage the improvement of vehicle fuel economy, such as individual country taxation options. In particular, reinforcing policies by EU member governments may well be required to encourage consumers to purchase the low fuel-consumption vehicles that ACEA is prepared to produce over the next 10 years. A recent analysis prepared for the UNFCCC Annex 1 Expert Group (Fulton 2000) estimates that, through 2008, as much as two-thirds of the CO_2-per-kilometre reduction achievable from technology uptake could be "lost" to increases in vehicle size, weight, and horsepower. For example, a 25 percent reduction in CO_2 per kilometre achieved through the use of new technologies could become a mere 8 to 10 percent reduction after accounting for consumer shifting to bigger, heavier, more powerful vehicles. Such a scenario, though pessimistic, would not be inconsistent with the types of consumer shifts that have occurred in recent years, especially in the United States, where minivans and sport-utility vehicles have become increasingly popular.

To support the agreement from the demand side and to counter unfavourable market shifts towards less fuel-efficient cars, the Commission is looking for a Community framework for vehicle taxation that allows tax-differentiation based on fuel efficiency. Also under development are a general framework and formats for consumer information (such as vehicle labelling) that would require the fuel efficiency of new cars to be more visibly displayed when vehicles are sold.

In contrast to the US approach, the current European strategy relies on relatively near-term, conventional technology progress and its incremental (but rather rapid) introduction into the new fleet. A radical change in car design to bring down fuel consumption rapidly is not envisaged. At the same time, for medium- and long-term car technologies, the EU research programme, Car of Tomorrow, is clearly less ambitious than its US equivalent, the Partnership for a New Generation of Vehicles (PNGV). As is the case with its US counterpart, the European research programme combines funds from different existing programmes ("Brite-Euram" for industrial and materials technologies; "Joule–Thermie" for energy research, development, and demonstration; ESPRIT for information technologies; and other programmes related to telematics and transport). Its initial set-up included detailed performance targets and research and development priorities, but the necessary collaboration and monitoring structure with industry has apparently been abandoned. Unlike its US counterpart, Car of Tomorrow now corresponds to a number of more or less independent projects that are not integrated into a process of progress evaluation and selection towards a common target (Commission Task Force 1996, EC 1997e).

What impact will the European voluntary agreement have on CO_2 emissions? In principle, its contribution could be very significant, and it may break the trend towards heavier and higher-powered cars. However, most of its effects on CO_2 emissions will be delayed until after 2010 because of the relatively slow rate of stock turnover, and especially because the car manufacturers foresee a slower increase in fuel efficiency in the early years of the agreement (until 2003) and only afterwards a quicker change to attain the target. With fleet renewal taking between 10 and 15 years and significantly more efficient new cars only entering the fleet after 2005, most of the CO_2 reductions will be "harvested" after 2010.

Conclusions

The European Commission, along with individual Member States, is concerned that the continuing increases in transport CO_2 emissions make it more and more difficult to achieve the Kyoto target. The Commission has prepared a number of strategy papers to tackle the greenhouse gas emission problem in transport. In most areas, it has to rely on Member States to act. A notable exception is the voluntary agreement with European car manufacturers, which is likely to have a significant impact in the reduction strategy. Several new Commission initiatives are imminent, such as the proposal for a Directive on complementary fuel specifications for the year 2005, the revision of the Common Transport Policy, a Green Paper on urban transport, and a Directive on the taxation of heavy goods vehicles for the use of certain infrastructure.

But in view of the recent upward trends in CO_2 emissions from the transport sector and the problem this creates for respecting the commitments undertaken by the EU in Kyoto, there may still be a strong need for reinforcement with additional policies and measures. While it wishes to underline the responsibility of Member States in establishing their own policies and measures to reduce CO_2 emissions, the Commission plans to reinforce Community-wide actions together with the development of an emissions trading system within the EU. Common and co-ordinated policies and measures could be strengthened by integrating environmental aspects into transport policies.

CHAPTER 9: THE UNITED STATES

Background

Trends in Transport and CO_2 Emissions

The United States has one of the most transport-intensive economies in the world, whether measured by travel and freight activity per capita or per unit of GDP. Relative to GDP, only Australia matches the US for travel. Reasons for such a high mobility in the US include a large land area, low travel costs, high incomes and decentralised land-use patterns. Widespread suburbanisation has resulted in increasing use of cars and other light-duty vehicles for relatively short trips.

The United States has the highest ownership rates of cars and of personal-use "light trucks" among IEA Member countries, the highest per-capita use of these vehicles, and by far the highest total travel both overall and by light-duty vehicles. In 1995, the average US resident travelled more than 23,000 kilometres, and 85 percent of those kilometres were travelled in private automobiles. In the same year, there were nearly 600 cars per 1,000 people (counting light trucks that are used for personal matters), or roughly 1.6 cars per household. There were more cars than licensed drivers.

Including light trucks, the US fleet of personal vehicles has the highest fuel intensity (energy use per passenger-kilometre) among the countries studied in this report. But it also showed the largest decline in fuel intensity during the 1970s and 1980s. If only cars are counted, the United States has an average energy intensity that is close to that of most European countries.

Table 9.1. Key Transport Statistics for the United States, 1997

Population (million)	**266**		
GDP per capita (US$ in PPP)	**30,514**	Cars (per 1,000 people)	**501 ***
Passenger transport activity (billion passenger-kilometres)	**6,360****	**Freight transport activity** (billion tonne-kilometres)	**4,842*****
Passenger car	*95.7%*	*Road*	*30%*
Bus and coach	*3.7%*	*Rail*	*41%*
Urban rail	*0.3%*	*Inland navigation*	*11%*
Railway	*0.3%*	*Oil pipeline*	*18%*
Transport final energy consumption (million tonnes of oil-equivalent)		**Transport CO_2 emissions** (millions tonnes)	
1990	**425.4**	1990	**1,221.2**
1997	**489.7**	1997	**1,409.7**
Rail	*2%*	*Rail*	*2%*
Road	*93%*	*Road*	*94%*
Inland navigation	*1%*	*Inland navigation*	*3%*
Pipeline	*4%*	*Pipeline*	*1%*

* Light trucks for private use (estimated at around 100 vehicles per 1,000 inhabitants) are not included.
** Domestic air (not included) accounted for 745 billion passenger-kilometres travelled in 1997.
*** 1996.
Source: US Department of Transport, OECD, IEA

Figure 9.1. CO_2 Emissions Trends in the United States, 1970-1997

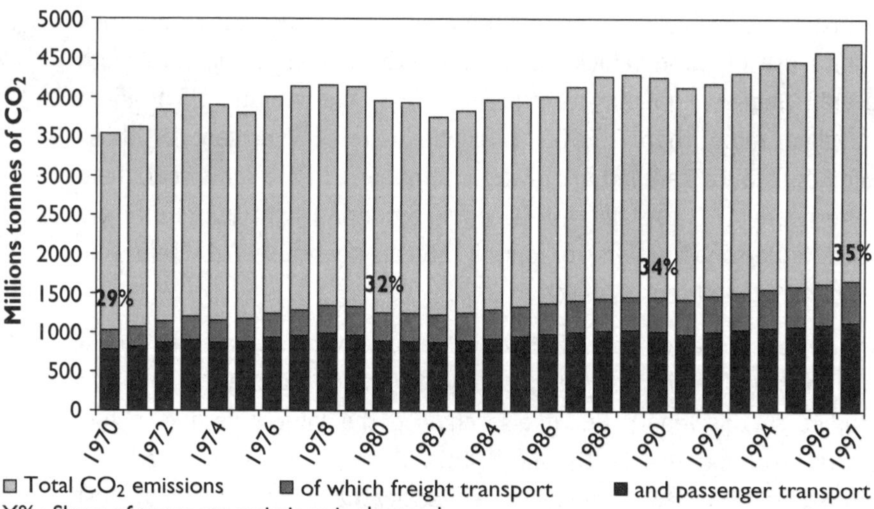

X% : Share of transport emissions in the total
Source: IEA/LBNL

Transport currently accounts for about 30 percent of US final energy consumption and about 32 percent of carbon dioxide emissions. Transport has been the second fastest growing sector in terms of energy consumption since 1990, after services. Between 1990 and 1994, transport energy consumption grew at an average rate of 0.7 percent per year, and 1.8 percent per year in the freight sector alone. This growth occurred despite real gains in transport-energy efficiency: Average fuel intensity in cars dropped from 17.8 to 11.6 litres per 100 kilometres, while truck fuel intensity dropped from 3.2 to 3.1 megajoules per tonne-kilometre, despite decreasing load factors.

In the United States, mass transit (urban buses and trains) accounts for only about 3 percent of total passenger travel. Thus, even a doubling of transit ridership would not reduce light-duty vehicle travel very much. Furthermore, both transit buses and trains have about the same energy intensity as cars; the intensity of light trucks is substantially higher.

In terms of freight travel, the United States is also different from most other countries. While the volume of freight per capita is second only to that of Canada among IEA countries, the modal mix features a higher share of rail and barge, relative to trucks, than in most IEA countries. US trucks are about average among IEA countries in energy intensity (fuel use per tonne-kilometre). So US energy use and emissions for freight, while high, are closer to levels in other countries than are the figures for tonne-kilometres of travel. And the Unites States is now getting more efficient in freight – during the 1990s, trucking fuel intensity fell steadily.

Figure 9.1 shows the trends in US CO_2 emissions from 1970 to 1997.

Public Attitudes and Perceptions

High mobility and motorisation in the United States are coupled with a generally higher degree of reliance on personal motor vehicles than in other countries. Many people in the United States appear to be unconcerned about the negative externalities of car use. At least they do not appear to incorporate such concerns into their transportation decision-making. This may reflect the fact that in many areas these externalities are low. Urban air quality is improving in most cities, as the Federal and state governments continue to press for stricter vehicle emissions requirements and to take other steps to improve ambient air quality. Governments are especially active in Clean Air Act "non-attainment" areas – that is, areas where measured levels of certain criteria pollutants exceed Federally-established guidelines.

Traffic congestion is an important concern, especially in the larger cities and their suburbs. And the response at the regional and local level continues to be more roadway capacity rather than dampening demand for vehicle travel. There is little support for measures to make driving more expensive, to develop alternative modes of travel, or to change land-use practices. This is not true everywhere; a few cities, such as Portland, Oregon, have in recent years adopted strongly pro-mass transit and anti-sprawl policies (Portland 1996).

US attitudes towards global climate change may also be different from those in many other countries. Polls suggest that US residents are confused about the science of climate change, relatively unconcerned about its effects during their lifetimes, and not very supportive of actions to mitigate it (see, for example, Gallup 1997). There is little public support in the United States for dramatic changes in transport policy to reduce greenhouse gas emissions from cars and light trucks. In the past, there *was* strong public support for the existing fuel economy standards, but there appears to be little support for new measures.

Role and Position of Industry

Although many industries will be affected by climate change and by the measures taken to mitigate it, two have played a particularly important role in the development of US transport-CO_2 policy-making: the automobile and oil industries. For example, these industries were well represented in the 1994-95 effort by the US Government and other parties to develop a greenhouse-gas emission-reduction strategy for personal motor vehicles (Dialogue 1995).

The auto industry has strongly resisted government influence on the types of cars and light trucks it sells. It has actively campaigned for freezing and even repealing existing standards (for example, CVC 2000). But the industry has also come out publicly in favour of higher gasoline taxes as a means to demand restraint (Dialogue 1995). More importantly, the car industry has strongly supported and participated in research and development efforts, such as the Partnership for a New Generation of Vehicles (PNGV). Recently both Ford and General Motors have committed to improving the fuel economy of some light duty trucks by up to 25% over the next few years.

The US car companies' interest in developing environmentally-friendly vehicles may also stem from a desire to meet California's increasingly strict requirements for low- and zero-emission vehicles. They realise that sooner or later the world of low gasoline taxes and non-changing fuel economy regulations can end, and they are concerned about foreign competition also moving in this direction. Recent introductions of low-emission, high-fuel economy hybrid (gasoline/electric) vehicles by Japanese companies may provide an additional spur to action. Finally, US manufacturers recognise the importance of selling into an increasingly global market. Providing vehicles with high fuel economy is an important part of being a "global player."

Notwithstanding their research and development efforts, US manufacturers have been slow to produce low fuel intensity (and low-CO_2) vehicles. As reflected in the PNGV agreement, the US industry's efforts appear to be focused on maintaining vehicle size and performance while decreasing environmental damages. The US auto industry believes that its basic product and the way it is used can remain nearly intact in an environmentally-constrained world. This contrasts with Europe and Japan, where the auto companies appear more willing to segment their product lines and offer very low fuel intensity cars for particular purposes or specific market niches (urban mini-cars and small gasoline/electric hybrids). These differences may also reflect differing travel and market conditions on the two continents. There are fewer constraints (such as lack of parking, narrow road, or taxes) on the purchase and use of large vehicles in the United States. As a result, there appears to be relatively little US demand for mini-cars or other performance-constrained vehicles.

Unlike the automobile industry, the oil industry remained until recently opposed to fuel or carbon taxes and actively lobbies against them, often citing residual doubts in the scientific community about the significance and impact of global climate change (see, for example, API 2000). Recently, however, some oil companies have re-assessed their stance and have adopted more conciliatory positions (BP Amoco 2000).

Other Influences and Trends

The US Energy Information Administration (EIA) provides an annual long-term forecast of energy use in US transportation. EIA's most recent forecast of road-transport CO_2 emissions through 2020 shows a 1.4 percent increase per year, surprisingly low considering that the drop in automotive fuel intensity over the past 20 years seems to have levelled off. One of the more uncertain aspects of transport modelling in the United States has been the issue of passenger travel saturation, both in car ownership rates and car use. US experts disagree as to whether the system as a whole is nearing saturation, with an accompanying levelling off of travel. Some argue that the current amount of travel per person is already near saturation and will not increase much more, especially in terms of hours per day spent travelling. Therefore, since the vast majority of passenger travel is already by car, total vehicle-kilometres of travel will begin to level off (Lave 1992). In addition, the retirement of the "baby boom generation" over the next 20 years will cause reductions in travel that will tend to offset increases in younger age groups. But some experts argue that continuing suburbanisation and increases in non-commuting travel distances will keep the growth rate in travel well above the population growth rate.

Evolution of Transport-CO_2 Policy

Post-World War II transportation policy in the United States was dominated by road construction. This was especially true after the passage in 1956 of the Federal-Aid Highway Act, which authorised expenditure of nearly US$2 billion per year for the continued construction of the national highway system (Wiener 1997). This programme was financed with newly instituted Federal gasoline taxes. For roads that qualified as Interstate Highways, the Federal government generally covered 90 percent of the costs, with the remainder coming from the individual states. Other roads eligible for Federal financing received up to 50 percent of construction costs from Washington. With money flowing so abundantly, all states embarked on ambitious road-building programmes.

Since the early 1990s, however, some Federal transport investments have gone to projects other than road-buildings because of several important changes in Federal transport policy. The most important of these were the Intermodal Surface Transportation Efficiency Act (ISTEA 1991) and its successor, the Transportation Equity Act for the 21st Century (TEA-21 1998). This new legislation has put more emphasis on intermodal solutions to metropolitan transport problems, allowing metropolitan areas to administer Federal funds themselves, rather than having to go through state transport departments. Nonetheless, due to the small share and energy-intensive nature of US public transportation, these increased funds may have only a marginal impact on US CO_2 emissions.

Another important change in Federal policy was the 1990 Clean Air Act Amendments. These amendments strengthened both the definition and the range of corrective actions available to areas when pollution is high. The United States Environmental Protection Agency (EPA) monitors the reporting of pollutant levels, as well as the progress of local action. Carrot-and-stick measures related to transport infrastructure funding enforce action in the non-attainment districts. So the Amended Clean Air Act is closely linked to ISTEA and TEA-21.

The third piece of legislation that made an important impact in the 1990s was the Energy Policy Act of 1992. Among other things, it authorised increased funding for alternative-fuel vehicles (AFVs) and initiated a larger programme to promote these vehicles. This programme requires governments and private fleets to purchase specified numbers of AFVs and it set targets for the displacement of oil by alternative fuels in the transport sector. By 1999, however, the displacement targets for 2000 and 2010 appeared unlikely to be met. Many of the provisions of the act, which require rulemakings by the Department of Energy and other agencies, have not been fully implemented. Alternative-fuel vehicles still represent much less than 1 percent of vehicle sales in the United States.

Various research programmes, including Intelligent Transportation Systems (ITS), the Partnership for a New Generation of Vehicles, and the Travel Model Improvement Program, have also received government attention and funding during the 1990s. A variety of forces, including environmental concerns, has driven many of these changes. But global climate change has not been an important force in US transportation policy and is unlikely to become one in the foreseeable future, although many policies are justified in part as having greenhouse-gas-reduction benefits in addition to their primary benefit.

Beyond, and within, these broad programmes, the United States has developed some transport policies and measures that are either focused on CO_2 reduction or have important implications for CO_2 emissions. Most notable of these were the policies included in the 1992 Climate Change Action Plan, which in turn were incorporated into the 1993 US National Communication to the UNFCCC. This plan contained four measures to reduce CO_2 emissions from transport:

- **Parking "Cashout".** This altered a provision of US tax law that prevented employers from offering employees a tax-free cash allowance in lieu of tax-free parking at the work place. Although put forward in the Action Plan in 1992, the measure was not enacted by Congress until the passage of TEA-21 in 1998. It is too early to tell the effectiveness of the legislation in increasing transit ridership.
- **Innovative Transportation Strategies.** This action consisted of seed money and incentives to state and local government for various pilot projects, including traffic control measures, congestion-pricing projects and other innovative strategies, particularly those focused on pricing of transport services.
- **Telecommuting Guidance.** This committed the government, through the EPA, to develop guidance for states and localities on how to earn

emission-reduction credits (under the Clean Air Act) by encouraging the adoption of programmes of private-sector telecommuting (i.e. when an employee is not working from his/her actual employer office, but from his/her home office). As of 1999, such guidance had not yet been issued.

- **Fuel Economy Labels for Tires.** This measure directed the Department of Transportation to develop a labelling system for new tires based on each tire's "rolling resistance" and its impact on fuel economy. It was estimated that using efficient tires would improve fuel economy by up to 4 percent. This measure faced such strong opposition in Congress that the Department of Transportation appropriations bill effectively forbade any analysis or further development of this measure, although a voluntary programme for heavy truck tires was developed.

It is unclear whether any of these four measures has had, or will have, a real impact on total CO_2 emissions. The 1993 National Communication estimated that, collectively, these programmes would save 6.6 million tonnes of carbon-equivalent by 2000, but this estimate was revised downward to 4.6 million tonnes in the 1997 National Communication. This is a small fraction of the total US greenhouse-gas-reduction commitment under the Kyoto Protocol, which is around 100 million tonnes.

Selected Transport-CO_2 Initiatives

CAFE

Corporate Average Fuel Economy (CAFE) standards were established in 1975 through the Energy Policy and Conservation Act. They took legal effect in the model-year 1978 for cars and in 1979 for light trucks. The CAFE standards require that the sales-weighted mile-per-gallon rating of new cars be above a certain standard. There are separate standards for cars and light trucks.

The EPA assigns a fuel-economy rating (calculated over the US City/Highway Fuel Economy Test drive cycle) for each engine family and vehicle model sold in the United States. Four separate CAFE ratings are calculated for each manufacturer on the basis of the sales-weighted fuel economy ratings across a manufacturer's domestic cars, imported cars, domestic light trucks and imported light trucks.

The CAFE rating for a particular manufacturer for a given year should be above the CAFE standard, after taking into account CAFE "credits" that are earned by achieving an above-standard average in any of the three previous and three following years. Failure to meet the standard results in a fine of US$5.50 per vehicle sold for each 0.1 mile per gallon the average falls below the standard. The CAFE standards for cars were set by legislation and can be changed through rulemaking by the National Highway Traffic Safety Administration (NHTSA). NHTSA has not changed the car standard since 1990. For light trucks, the standard is set by NHTSA for each model year. In recent years, Congress has expressly forbidden NHTSA to set the light truck standard at any level other than that of the previous year (APP 1998).

Figure 9.2. CAFE: Standard and Realisation in the US, 1978-1999

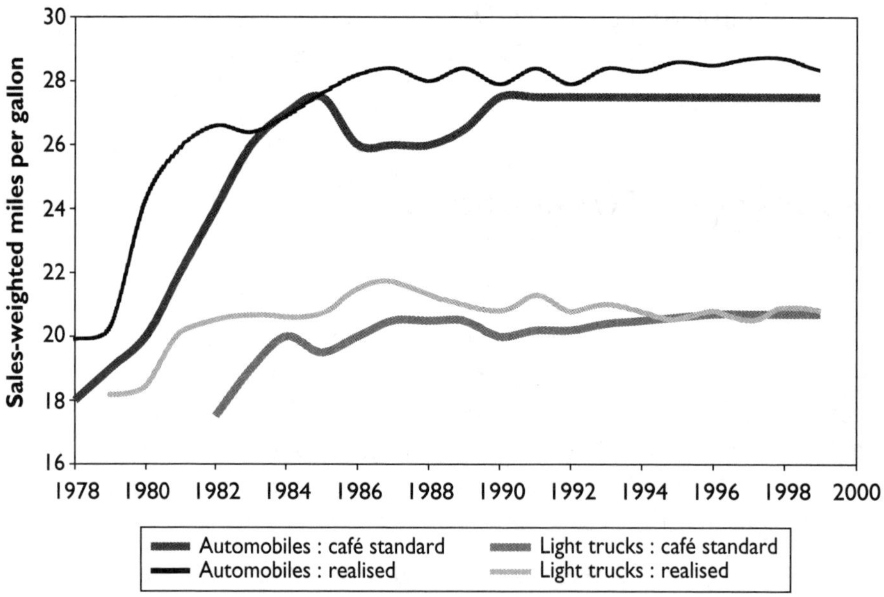

Source: ORNL 1999

The CAFE law also provides a system of credits for the production of alternative- and flexible-fuel vehicles. (A flexible fuel vehicle can handle any mix of gasoline and alcohol fuel up to 85% alcohol). Electric and dedicated alternative-fuel vehicles are counted as very high mpg vehicles in the CAFE average. Dual-fuel and flexible-fuel vehicles are also given additional mpg value, although their contribution cannot increase a manufacturer's CAFE rating by more the 1.2 miles per gallon.

A complementary programme of vehicle taxation, known as the "gas-guzzler tax," is in place for cars that do not meet certain minimum fuel economy standards. It does not apply to light trucks. The annual system of gas-guzzler fees has been adjusted several times. In 1980, it covered vehicles that got 15 miles per gallon (15.6 litres per 100 kilometres) or less with fees up to US$550. By 1991, it covered vehicles that got 22.5 miles per gallon (10.5 litres per 100 kilometres) or less with fees from US$1,000 up to US$7,700 (ORNL 1999).

By the end of 1997, total fines from the CAFE system amounted to US$559 million in 1997 dollars (ORNL 1999). Annual tax receipts from gas-guzzler fees rose from just over US$1 million in 1980 to US$164 million in 1992, then declined to US$48 million in 1997, totalling over US$1.6 billion for the whole period.

Changes in the fuel economy of cars and light-duty trucks are shown in Figure 9.2. The following key trends are worth noting:

- Between 1978 and the mid-1980s, there were significant improvements in fuel economy for both cars and light trucks.
- From 1986 till now, fuel economy for both cars and light trucks has remained nearly constant.
- The rising share of light trucks has caused the average fuel economy of all light-duty vehicles to fall since the late 1980s.

Light-duty truck sales have increased from about 20 percent of the market in the early 1980s to 46 percent in 1998. They now consume more fuel than cars do. This rise in light truck sales is the primary reason for the 2.1 miles per gallon decline in fuel economy for new light-duty vehicles from its peak in 1988 and its 1998 level (EPA 1999).

Curb weights and engine sizes fell significantly between 1975 and 1980 and continued to fall until the mid-1980s. Over the past 10 years, however, dramatic increases in the average weight of vehicles and in average engine size were key offsetting factors preventing additional improvement in the fuel economy that would have occurred through technical improvements. The average vehicle interior volume has remained remarkably stable through the 1980s and 1990s. But horsepower-to-weight ratio has increased dramatically since the mid-1980s and is now 50 percent higher than in 1980. EPA estimates that fully 5 miles per gallon has been "traded" for improvements in other attributes since 1986. In other words, if vehicles had been held at their horsepowers and weights, average fuel economy in 1998 would have been 5 miles per gallon higher than it actually was.

Although steep increases in the CAFE standards coincided with equally sharp increases in fuel economy from 1978 to 1986, it is not clear that the two phenomenon were closely related. In 1979, fuel prices also began climbing sharply, just as the standards came into effect. Consumers were faced with a spike in gasoline prices and had an economic incentive to choose more fuel-efficient cars. While prices eventually came back down – precipitously after 1986 – technical gains nevertheless had been made in car design and manufacturing which did not go away with lower prices. For this reason, different opinions exist regarding the effectiveness of the CAFE standards. Some analysts have argued that CAFE has had a significant impact (Greene 1998), but others have attributed progress to high prices, limiting CAFE benefits to periods when fuel prices were low (for example, Nivola and Crandall 1995).

Partnership for a New Generation of Vehicles

The US Partnership for a New Generation of Vehicles (PNGV) programme is a partnership among 11 US Government agencies and the US Council for Automotive Research, a co-operative research effort of the three major US auto companies. The programme is intended to "develop commercially viable vehicle technology that, over the long term, can preserve personal mobility, reduce the impact of cars and light trucks on the environment, and reduce US dependency on foreign oil" (DOE 1999a). The programme has as a principal goal the development of a vehicle "with triple the fuel efficiency of today's mid-size cars while maintaining or improving safety, performance, emissions and price" (DOE 1999b). The term "today's vehicles" refers to 1994 model year vehicles (PNGV 2000). The timetable is to develop "concept" vehicles at each car company by 2000 and a production prototype vehicle (that is, one that could be mass-produced within three to five years) by 2004.

While the PNGV programme involves a considerable research and development expenditure by the US government (reported as US$250 million in 1999), the amount of investment by industry itself is reported to be close to US$1 billion per year (CBS 2000).

A key aspect of the agreement is the maintenance of safety and price at current levels. Vehicles getting three times the fuel economy of 1994 average vehicles can be made without any technological breakthroughs, but maintaining these attributes, especially price, represents a daunting challenge. Another challenge for manufacturers may be to follow through with marketing these vehicles if they prove successful as prototypes. There will certainly be a temptation to use the technologies developed for the prototypes to produce larger, more powerful vehicles with something less than tripled fuel economy, rather than 80-mile-per-gallon (3 litre per 100 kilometre) vehicles with 1994 levels of performance.

Perhaps in order to address these types of concerns, a proposal to encourage market adoption of advanced technology vehicles was recently put forward as part of the current administration's Climate Technology Initiative package of budget proposals for the fiscal year 2001. It includes the following specific tax credits for the purchase of high efficiency vehicles:

- **Tax Credit for Electric Vehicles and Fuel Cell Vehicles.** There is an existing 10 percent tax credit, subject to a US$4,000 cap, for the purchase of qualified electric vehicles and fuel cell vehicles. This tax credit is currently slated to phase down in 2002 and be eliminated by 2005. The proposal would extend the credit at its full level through 2006.

- **Tax Credit for Highly Fuel-Efficient Hybrid Vehicles.** The proposal would provide a new tax credit of US$1,000 for hybrid vehicles, including cars, minivans, sport utility vehicles and pickup trucks, purchased in 2003-2004 and that are at least one-third more fuel efficient than a comparable vehicle in the same class. Similarly it would provide US$2,000 for hybrid vehicles from 2003-2006 that are at least two-thirds more efficient; US$3,000 in 2004-2006 for hybrid vehicles that are at least twice as efficient; and US$4,000 in 2004-2006 for hybrid vehicles that are at least three times as efficient.

While the US$4,000 credit for Electric Vehicles apparently has not been large enough so far to spur significant sales in the US, the market-adoption incentives proposed for hybrid vehicles could provide an important boost to the prospects of these vehicles. Similar incentives for other advanced technologies, such as fuel cells, could represent an important complement to the PNGV policy already in place.

California Low-Emission Vehicle Program

The California Air Resources Board (CARB) has developed a Low-Emissions Vehicle (LEV) programme that aggressively seeks to promote the introduction of low-emission vehicles over the period 1990-2010 (CARB 2000). Although it is focused on air quality, the programme may have profound impacts on the market success of new engine technologies and alternative-fuel vehicles, which in turn could have a major impact on the fuel economy of future vehicles and, thus, on CO_2 emissions.

The LEV programme was begun in 1990. It originally required that, by 1998, 10 percent of all new vehicles sold in California were to be low-emission vehicles and that 2 percent needed to be zero-emission vehicles (ZEV). These proportions were to reach 20 percent and 10 percent, respectively, by 2003. In 1996, however, CARB abandoned the 1998 deadline, while retaining the 10 percent ZEV sales requirement for 2003. CARB recently adopted "LEV II," which sets out programme guidelines and requirements through 2010. Partial ZEV "credits" will be available for "super-ultra-low emission vehicles" (SULEVs), a new category of vehicles that produces near-zero emissions. Some fuel-cell and electric-hybrid vehicles may qualify for this category.

The ZEV mandates have been important in the United States for several reasons. First, they appear to have had an impact on the major car manufacturers' timetables for developing and marketing low-emission vehicles. Most vehicle manufacturers now produce one or more gasoline models that qualify as "LEV". Five years ago, there was considerable debate as to whether gasoline vehicles would ever be able to qualify in this category. The prospect that other states might adopt similar mandates, particularly in the Northeast, probably played an important role in acceptance by the auto industry of EPA's "Tier II" regulatory process, which is likely to result in significantly tighter national emission requirements for both vehicles and gasoline by 2004.

While car companies have shown that they can certify conventional gasoline vehicles as LEV and even ULEV (ultra-low-emission vehicles), it appears unlikely that these vehicles can be certified as SULEV or ZEV. After 2002, therefore, companies will need to produce advanced-technology vehicles, like those being developed and demonstrated in the PNGV programme, or electric vehicles. In order to sell vehicles in California, these high efficiency cars must make up 10 percent of overall sales. With more than a million vehicles a year sold in the state, the requirement could single-handedly spur large-scale production of next-generation and alternative-fuel vehicles in the United States.

ISTEA/TEA-21

Although they are clearly focussed on other matters, the Intermodal Surface Transportation Efficiency Act of 1991 and its successor, the Transportation Equity Act for the 21st Century (TEA-21), could have important impacts on CO_2 emissions. TEA-21, passed in 1998, is the largest public spending bill in the history of the United States. Most, but not all, of the funding comes from Federal gasoline taxes. The Act authorises an estimated US$217 billion to be spent on transport over six years, an average of US$36 billion per year (FHWA 1998). About 78 ercent of those funds are for "Title I" projects – that is, projects carried out under Federal Highway programmes. About 20 percent goes towards "Title III" projects – those that are formally administered by the Federal Transit Administration. Much of this money will probably be used to help metropolitan areas improve their public transport systems, which have been losing passengers since the 1920s. Whether such actions serve to reduce CO_2 emissions at all is unclear, given that public transit has about the same fuel intensity as automobiles. More promising are TEA-21's provisions to allow local governments and metropolitan planning organisations to use transport money on land-use or urban-planning projects.

Another aspect of TEA-21 that might affect CO_2 emissions is the "Value Pricing" programme, which replaces the Congestion Pricing Pilot programme under ISTEA. The earlier programme provided seed money for up to five demonstration projects of road pricing in congested corridors in the United States. Under ISTEA, no project got off the ground, however, because of local politics and opposition. A small change under the new programme is that states can price Interstate Highways that were constructed using Highway Trust Fund money, something that had been previously forbidden. It is hard to tell what the cumulative effect of these TEA-21 measures on greenhouse gas emissions will be, but their overall impact is likely to be small. Nonetheless, some of the individual programmes now being funded may provide clues as to how to proceed with transport policy in ways that will lower greenhouse gases and improve sustainability more generally.

Speed Limits

Another Federal transport policy that can affect energy efficiency and CO_2 emissions is speed limits.

Before 1974, speed limits were a state prerogative. In 1974, during the "energy crisis," a 55-mile-per-hour (about 90 kilometres per hour) Federal limit was established as a means of reducing energy consumption and improving safety. In 1987, the law was changed to permit states to raise the speed limit on rural interstate highways to as high as 65 miles per hour. In 1993 this was extended to certain non-interstate highways.

Finally, in 1995, the National Maximum Speed Limit law was repealed, allowing states to set their own speed limits with no upper bound. A few states, such as Montana, now have roads with *no* speed limits (only recommended maximum speeds).

By the time the Federally-mandated speed limit was raised and then abandoned, arguments against raising speed limits were no longer focused on energy, but on the impact on traffic fatalities and injuries. It is hard to know precisely what impact these changes have had on energy consumption, but unpublished Department of Energy calculations estimate that the impact on US vehicle fuel use is about 150,000 barrels per day. EPA separately estimated that the change could cause CO_2 emissions to rise between 6 and 15 megatonnes per year (about 6 to 15 percent of the required US CO_2 reductions under the Kyoto Protocol).

Research and Development Initiatives

Various research and development initiatives are underway, sponsored or assisted by the Federal government. Some of these focus on automotive technology, such as the PNGV and various projects studying alternative fuels. Moreover, technologies developed as a result of other research sponsored by the Federal government, including military and space research, can find their way into automobile technology.

The Federal government is also spending a great deal of resources on system- and operations-related research, including the Intelligent Transportation Systems Programme, and the Travel Model Improvement Programme. The former is designed to develop technology to allow the existing infrastructure to accommodate increased demand. Satellite-assisted computer technology would manage highway, rail, and air traffic both to allow for smoother operation and increase capacity without actually deploying further infrastructure. Advocates of Intelligent Transportation Systems argue that, since the technology can help reduce congestion along networks, vehicles will be able to operate using optimal efficiency. Transport economists counter that if congestion reduction lowers the cost of travel, it may induce more traffic, and that the emissions caused by the new traffic might more than offset the efficiency gains (Fulton et al. 2000). The Travel Model Improvement Programme is a long-term project designed to increase the accuracy of metropolitan-wide transport models used in the transport planning process. While this programme has no direct effect on energy consumption, the models developed may be better able to reflect the impact of various policies – including land-use and other environmentally-oriented measures – than currently available models.

Alternative Fuels

The 1992 Energy Policy Act provided specific goals for the use of alternative fuels in the transport sector: 10 percent of light-duty gasoline was to be displaced by 2000, and 30 percent by 2010. However, as of 1998 only a very small percentage of vehicles on the road were Alternative Fuels Vehicles (AFVs), about 300,000 out of over 100 million. Nonetheless, the rate of increase in the 1990s was high, with much of it attributable to the Energy Policy Act. The 2000 displacement goal may be met. But if is, the cause will be the increasing use of non-gasoline blends in gasoline as a fuel in conventional gasoline vehicles, not the use of alternative fuels in alternative-fuel vehicles. It now appears unlikely that the 2010 goal will be met without additional policies and measures.

In addition, it is unclear whether meeting such a goal would provide large reductions in CO_2 emissions. It would depend on *which* alternative fuels were adopted. An analysis by the US Department of Energy concluded that the widespread use of alternative fuels by 2010 might have little impact on either CO_2 emissions or US dependence on energy imports (DOE 1995).

Historically, three disparate factors have influenced the push towards AFVs:

- Air-quality problems in cities like Los Angeles, New York, and Washington, D.C., have made reductions in nitrogen oxides, non-methane hydrocarbons, and carbon monoxide a priority in transport policy. Some alternative fuels provide reductions in these pollutants.

- Fears for US energy security, dating back to the oil crises of the 1970s, have helped spur the quest for domestic sources of energy, particularly for the transport sector. Some alternative fuels can be produced domestically.
- Policies promoting biofuels have played a role in US agriculture policy, as a way to ensure a market for farm crops. The US policy of subsidising the price of ethanol has received strong support from Midwestern states.

These factors spurred a fairly strong, but "fuel-neutral," research and development programme, which was adopted both by Congress and supported by each recent Administration. But a reluctance to try to "pick winners" by favouring a particular alternative fuel has prompted multiple programmes and demonstration projects, focused on different fuels, with no single fuel emerging as dominant. At the same time, there has been little relationship between programme spending on different fuels and the greenhouse-gas reduction potential of each fuel. The greenhouse-gas reduction benefits of different alternative fuels vary widely, with some fuels (grain ethanol, natural gas and methanol) having relatively little impact, while others (cellulosic ethanol, bio-methanol, and fuel cell vehicles) could dramatically reduce emissions per vehicle (IEA 1999).

Cellulosic ethanol could reduce CO_2 emissions per kilometre by more than 80 percent compared to gasoline (in similar vehicles), but it is now uncompetitive with grain ethanol (which accounts for about 1.6 percent of total US vehicle-fuel use). If advances expected by the US Department of Energy are achieved, it could be possible to produce 40 billion litres of cellulosic ethanol per year and use all of it as a gasoline blend in today's conventional vehicles. If, in the long term, cellulosic ethanol were expanded to 20 billion or 40 billion litres per year, substantial numbers of ethanol/gasoline flexible-fuel vehicles would be required. Ethanol could then play a significant role in reducing transport-sector emissions. Consequently, policies to encourage such low-emission fuels could have an important role in a long-term US strategy to reduce greenhouse gas emissions.

Conclusions

Transport-CO_2 policies in the United States are heavily focused on technological solutions. Changes to the fuel intensity of cars and light trucks, and to a lesser degree to the types of fuels consumed, make up the bulk of US transport strategy for CO_2 emission reduction. This technological focus appears to stem from a political and social aversion to policies aimed at behavioural change, such as significant increases in fuel taxes. This orientation has rendered US CO_2 policy significantly different from that of other industrialised countries, particularly those in Europe.

The US approach has produced some dramatic success stories. In the past 30 years, for example, ambient air quality has improved in many US cities, in large part because of technical improvements to new cars (US EPA 1996). Whether a purely technological approach to the problem of reducing greenhouse gas emission can work as well, remains to be seen. While the CAFE law has been the centrepiece of US climate change policy in recent years, the PNGV programme appears to be the cornerstone of future CO_2 policy. As a voluntary programme, however, this policy focuses on technical solutions in lieu of asking Americans to change their lifestyles. It also provides an unclear mechanism for ensuring that those solutions are achieved. Manufacturers appear committed to follow through with their commitments to build production prototypes, but it remains to be seen how many PNGV-style vehicles will in fact be produced and sold in the United States, and over what time-frame. The recent administration proposals to implement tax incentives for the purchase of PNGV-type vehicles represents an important, but as yet unimplemented, step in building markets for vehicles developed under this programme.

There do not appear to be significant opportunities in the United States to shift passenger transport away from energy-intensive modes of travel to less energy-intensive modes, simply because such modes do not appear to exist in the United States. On average, a city bus today emits about as much CO_2 per passenger-kilometre as a car. The energy intensity of urban buses has doubled since the 1970s, reflecting declining ridership rates.

Hence, mode-shift solutions to reduce the energy intensity of passenger transport will only work if they increase ridership on existing transit systems. Policies that increase transit capacity will probably not yield CO_2 benefits.

The United States is also exploring the use of low-emission fuels as a climate-change strategy, most notably in the development of cellulosic ethanol technology. If such a strategy could be successfully pursued, substantial emission reductions could be achieved with little impact on consumers, because the fuel could be blended into gasoline or used in fuel-flexible vehicles that are indistinguishable from conventional vehicles. While this strategy has a real potential to reduce emissions, cost barriers should be overcome. Current research efforts are unlikely to lead to commercialisation of this technology in the near future. In addition, purchase subsidies may be required to establish a market.

SUMMARY AND CONCLUSIONS

Key Findings

Since the signing of the Kyoto Protocol in 1997, carbon emissions from transportation have increased in virtually every IEA country. In some countries, growth rates in automobile use and freight transport show signs of slowing, but in others there are no such signs. Even with improvements in energy intensity, rising economic activity has led to significant increases in transport CO_2 emissions in every country studied here. In this environment, Parties to the Protocol face the formidable task of slowing the growth in emissions and then turning them downward. From the review of each country's policies and programmes, we report a number of conclusions.

Each country has identified significant potential for reducing or restraining transport emissions, but few expect to achieve reductions sufficient to offset the growth of transport activity in the near term. Among the countries that set *transport-specific* goals in the mid to late 1990s (the UK, the Netherlands, Denmark, Sweden), none expects to attain these goals by 2010. In the near term, policy makers have focused on the much more modest objectives of *reducing emission growth rates* from transport below "business-as-usual" trends. Many policy makers believe it will take decades for CO_2 emissions in the transport sector to stop growing and start to decline. Nonetheless, government policy intervention is important to avoid even higher emissions.

All current policies on carbon emissions from transport are complicated and hard to put quickly into operation. This is true for approaches which focus explicitly on CO_2 and for those with a wider environmental focus which achieve emissions reductions indirectly. No country has yet managed to put in place more than a small number of the measures that were under consideration in the mid-1990s. Much of the lag can be attributed to slow political processes and to political ambivalence. But even when approved, such policies can take several years to implement.

Table 10.1. CO_2-Transport Policy Classification in the Countries Studied

	Policies to change fuel prices		**Policies to change other variable costs**	**Policies to influence traffic flow**		**Policies to reduce demand for personal vehicle travel**	**Policies to enhance public transit**	**Policies to influence personal vehicle demand**
EXAMPLES OF POLICIES	*Fuel Tax*	*Carbon Tax*	*Road pricing; parking charges; varying other costs*	*Capacity expansion Intelligent Traffic System, computerised traffic management*	*Enhancing vehicle on-road efficiency (driver training, vehicle on-board diagnostic equipment speed limits, vehicle maintenance)*	*Telecommuting, Traffic Demand Management strategies, traffic calming*	*Expansion of service to reduce travel time and wait time, expansion of service area, enhancement of comfort, reduction of transit fares by direct (public sector) investment or regulatory reform to encourage private sector investment*	*Targeted acquisition, ownership, and registration fees; taxing company car benefits as ordinary income; revenue-neutral cross-subsidy schemes, such as feebates*
DENMARK	Fuel tax increases to dampen private consumption 1995-2002 Repeal of diesel tax refunds for truckers	Small yearly CO_2 tax increase has been put into place	New taxes on ticket purchases for air travel put in place in 1997				Lowering of prices for mass transit from May 1997	Very high ad valorem vehicle purchase tax, with a fuel consumption component added in 1999 providing lower taxes for very low fuel consumption cars; Green ownership fee modified to include fuel consumption component in 1998; Truck purchase taxes differentiated according to truck size
GERMANY	Fuel tax increase (30% / 20% for gasol/ diesel since 1990; Further increases envisaged			Near-term market introduction of ITS (traffic management, traffic information systems)			Most transit decisions have been devolved to state and local governments, although federal government funding for transit infrastructure has been increased	Vehicle ownership tax proportional to engine displacement, includes pollutant and some CO_2 emission differentiation

Policies to influence vehicle supply/production	Policies to influence urban structure	Policies to influence public attitudes toward transport and fuel consumption	Policies for promoting fuels low in greenhouse gas and alternative fuels	Policies to reduce CO_2 emissions from freight transport	
Advancing vehicle fuel economy or alternative fuel technology via assistance with manufacturer research and development, sponsorship of independent R & D, regulations and manufacturer performance mandates and tradable permits	*Land-use controls for compact urban development; co-ordination of transportation and land-use development; tax benefits to firms and households choosing to locate in "accessible" parts of metropolitan regions; enterprise zones; location-efficient mortgages*	*Media campaigns, youth education campaigns, information "exchange" projects, vehicle fuel-consumption labelling, training and education in fuel efficient driving etc.*	*Research programmes, support for development of fuel and vehicle infrastructure, fuel subsidies, vehicle purchase requirements*	*Increased trucking efficiency (higher load factors, shorter routes, fewer empty backhauls), mode switch from truck to rail or water, changes in overall demand for freight haulage*	**EXAMPLES OF POLICIES**
Following ACEA/EU voluntary agreement. Not other specific Danish policy	"Transport aware planning" of building development and transport infrastructure; proposal to limit growth to already built areas and restrict out-of-town shopping centres not yet enacted	Enactment of a new-car fuel economy labelling law (March 2000)	Current policies promote electric vehicles; proposal to encourage development of biofuels, principally biodiesel and ethanol		**DENMARK**
VA with German manufacturers, minus 25% fuel consumption between 1990 and 2005				Rail infrastructure financing, privatisation reform, rail freight/combined transport platforms, funding for construction of 52 new rail depots to facilitate intermodal transfers with trucks	**GERMANY**

	Policies to change fuel prices		**Policies to change other variable costs**	**Policies to influence traffic flow**		**Policies to reduce demand for personal vehicle travel**	**Policies to enhance public transit**	**Policies to influence personal vehicle demand**
THE NETHERLANDS	1997 fuel price increases as part of move toward increased variability of taxes		Overarching goal to switch fixed costs to variable costs for both cars and trucks; began with fuel price increases and yearly fee decreases in 1996; road pricing between cities announced for 2002		1999 measure to encourage use of econometer, cruise control, and on-board computer in cars; increased speed limit enforcement			1999 measure to encourage use of econometers, cruise control, and on-board computers in cars; increased speed limit enforcement
SWEDEN	SEK 0.25 (€ 0.3) increase in 1997; slight increase annually to keep pace with inflation; lower taxes on unleaded gasoline and low-sulphur diesel	Carbon tax component represents about 20% of gasoline fuel tax; Carbon tax on domestic aviation fuel rescinded in 1997	New limits on tax benefits of using a company car					Environmental impact differentiation in Vehicle purchase tax; changes in reimbursement for use of own car for business purposes that will now disfavor using larger cars
UNITED KINGDOM	Fuel cost escalator established in 1993 that annually increased fuel duties by 6% above inflation; switched to annual review basis in 1999		Local governments authorised to enact road pricing schemes as part of broader package if they so choose			Road Traffic Reduction Act (1997) directs local authorities to estimate existing and anticipated future levels of traffic, designate a reduction target and enumerate measures to reach the target		CO_2-based vehicle excise duty recently announced, to begin in 2001 (duty on new cars ranges from £90 to £160 depending on fuel type and CO_2 emissions per km)

Policies to influence vehicle supply/production	Policies to influence urban structure	Policies to influence public attitudes toward transport and fuel consumption	Policies for promoting fuels low in greenhouse gas and alternative fuels	Policies to reduce CO_2 emissions from freight transport	
	1993 "ABC" plan implemented to encourage location of major employment and residential centres near mass transit; reduce employer-provided parking, clustering of settlements; more bike paths			Plan in place for shifting some freight from trucks to rail and or barge, through development of intermodal facilities, greater truck cost variability, and other infrastructure improvements	**THE NETHERLANDS**
Volvo VA 1996 for 25% reduction in new-car fuel intensity; Pioneer of differentiated new-vehicle and yearly fees to favor cleaner engines and fuels			Carbon-tax component to fuel taxes that promotes low Carbon fuels		**SWEDEN**
Clean Vehicles Task Force formed to further develop programmes to promote production of low-emission cars	Policy guidance documents by national government to help local authorities in land-use and facilities-siting decisions, which provide clear preference for city-centre locations; accessibility to public transport becoming important land-use criterion as well	1997 transport White-Paper raised general awareness about transport issues and policy; Best-practices programme promotes and publicises energy-efficient practices; government supports local efforts aimed at travel reduction (HeadStart, TravelWise, Green Commuting Plans)			**UNITED KINGDOM**

	Policies to change fuel prices		Policies to change other variable costs	Policies to influence traffic flow		Policies to reduce demand for personal vehicle travel	Policies to enhance public transit	Policies to influence personal vehicle demand
EUROPEAN UNION	Increase in minimum fuel tax requirements for member States (but all States currently tax well over this minimum)		Ongoing efforts to harmonise road and rail regulations and prices					
UNITED STATES	US$ 0.15 per litre subsidy of farm-based ethanol		Tax law has been modified to permit parking "cashout" programmes; Congestion Pricing Pilot Programme to provide funding for pilot projects	Significant investment in Intelligent Transport System (ITS) technologies Many ITS projects in various stages of implementation in different cities Under the Transportation Equity Act (TEA-21), a significant amount of road-building projects	Eliminated national speed limit, 1995		Significant amount of research and demonstration project money under TEA-21; financial assistance to many transit agencies; subsidies to Amtrak, national passenger rail company	

Policies to influence vehicle supply/production	Policies to influence urban structure	Policies to influence public attitudes toward transport and fuel consumption	Policies for promoting fuels low in greenhouse gas and alternative fuels	Policies to reduce CO_2 emissions from freight transport	
Voluntary agreement with European Manufacturers association, goal of 140g CO_2/km by 2008					**EUROPEAN UNION**
Corporate Average Fuel Economy (CAFE) standards still in effect but required levels have not changed significantly in over 10 years; Gas-guzzler tax rates raised dramatically in early 1990s but revenues now declining as few vehicles are still subject to it; Partnership for a New Generation of Vehicles (PNGV) programme provides R&D and targets for development of prototype very low-fuel intensity autos		Fuel economy labelling	Energy Policy Act authorises several programmes to promote alternative-fuel vehicles (AFVs), including government AFV purchase requirements, tax credits for purchase of electric vehicles; Department Of Energy Clean Cities programme helps cities establish refuelling infrastructure, among other things		**UNITED STATES**

As reflected in Table 10.1, a majority of countries have linked CO_2 policies directly to comprehensive transport policy reform, with an emphasis on "getting the prices right." Other measures have focused on reducing the growth rate (or outright levels) of personal vehicle travel and improving traffic conditions. These policies typically are driven more by congestion and nuisance concerns than CO_2 reduction concerns. Only technology policy to reduce fuel intensity and promote alternative fuels is relatively independent. In practice, transport policy reform includes wide-spread reforms in the way transport services are priced and taxed, internalisation of environmental costs and integration of the infrastructure across modes. These measures, if broadly and vigorously enacted, might lead to 5-10 percent lower emissions than otherwise by 2010, mainly by reducing the distance driven in cars and trucks.

The overarching element of transport reform that links CO_2 measures to others is pricing. An important EU Green Paper in the mid-1990s supported a strong and central role for pricing. The plan includes internalisation of costs, including that of CO_2 emissions, as well as switching transport-related fees and taxes from a fixed to a variable basis where possible. Higher fuel prices, road pricing for all vehicles and pricing of freight haulage by the tonne-kilometres hauled are all included in broad strategies developed by individual countries. Raising the acquisition fees or yearly taxes on vehicles, especially those with high fuel-consumption is another favoured measure. So far, Denmark has made the strongest move towards a fuel-consumption based fee system where both new and existing vehicles are taxed in part on their fuel consumption rather than vehicle weight or engine size.

The high level of road-fuel taxes in Europe means that most carbon taxes by themselves will have only a modest impact on vehicle use and vehicle choice. For the US the relative importance of even a small carbon tax would be larger because of the low price of fuels there. Still, even small carbon taxes are an important element that helps distinguish the prices of high- and low-carbon fuels. Taxes also limit the impact of the rebound effect where more efficient vehicles can end up increasing car use. Still longer-term, foreseeable fuel price increases are expected to play a role. The UK has had a strong yearly tax-increase scheme. Sweden and

Denmark have enacted more modest yearly increases. The Netherlands and Denmark have announced schedules of price increases, with Germany now on the same track.

The most important technology policy measure now in place is the voluntary agreement for emissions reductions in new cars, between the European Union and ACEA, the European Association of Car Manufacturers. Japan and Korea have a similar agreement with their manufacturers. This EU-ACEA agreement will account for the bulk of reductions from baseline we expect for the European countries we studied. If successful, this measure alone could reduce emissions from cars by 15 percent to 20 percent below trend by 2010, with even greater reductions after 2010 when the more efficient new cars will have fully penetrated into the vehicle stock.

The goal of the voluntary agreement in Europe will be sought through advanced conventional technologies, mostly diesel and gasoline direct injection. "Next generation" engine types, hybrids and fuel cells, have been the subject of considerable private and government research efforts in a number of countries. They will be appearing on the market in small quantities during the coming decade but are not likely to play a major role until after 2010. Although hybrids are already for sale in the US, their emergence appears to be partly due to Californian pollutant regulation. Sales are unlikely to have a real effect on average fuel economy for many years. During the coming decade and beyond, government action may be necessary to promote their introduction. These technologies must overcome high initial costs and market inertia and achieve high levels of production before they can compete on their own merits.

The largest CO_2-focused research in the public sector is the US Partnership for a New Generation of Vehicles (PNGV). PNGV may well represent the most ambitious – but also most uncertain – of the elements in the current US basket of transport, energy and CO_2 policies. Detroit can already build cars "with triple the fuel efficiency of today's (1994) mid-size cars," with a number of prototypes already demonstrated by mid-2000. But whether auto makers can produce vehicles in the short run that are highly fuel efficient, while maintaining or improving safety, performance, emissions and price remains to be seen. Furthermore, once production prototypes have been built, it is not certain that car companies will market them aggressively. There will certainly be a temptation for the manufacturers to use the technologies developed to produce larger, more powerful vehicles with something less than maximum fuel economy, rather than 80 miles per gallon vehicles with ten-year-old levels of performance. Therefore complementary policies, such as those recently proposed in the US to provide purchase incentives for PNGV-type vehicles, may be important to ensure that consumers become interested in purchasing these vehicles, and that manufacturers produce them with fuel economy levels that meet the conditions for the incentives.

Measures to encourage modal shifts and non-motorised transport modes, and measures to encourage less personal travel, are part of all the European programmes, but it is difficult to judge what savings they have realised or will realise on their own. Improved transport services are expected from the privatisation of railways, but the ultimate effect is unclear. Land-use planning forms will complement transport policy reforms, but it is also difficult to predict how much this will reduce travel or induce modal shifting. In the European countries studied, infrastructure expansion and investment is increasingly tilting towards non-road or inter-modal facilities. At the minimum, these efforts will slow the erosion of the share of low-CO_2 modes.

The countries with the boldest and most comprehensive plans have either had to modify them to gain political acceptance or to withdraw them after they were accepted because of difficulties in implementation. Transport systems have already been tinkered with by government policy to the point where any additional pressure is strongly resisted by one or more involved group. More fuel tax increases no longer look viable in the near term and other approaches to raising the marginal cost of travel are also met with fierce resistance. Technology-oriented solutions, generally much more popular with the public, often require long-lead times to make real changes in vehicles on the road, and to bring down costs.

Each government has experienced minor setbacks in political tests, or has been forced to weaken policies or goals. Even attempts to make what appear to be "minor" policy adjustments are sometimes unsuccessful, such as the effort to implement a new tyre labelling policy in the US The Netherlands needed several years to pass a modest tax relief measure to encourage the purchase of econometers and other instrumentation on new cars. The UK government had to soften its regular fuel tax increases. Sweden and Denmark scaled back the fuel-economy goals they proposed in the mid 1990s when the less ambitious EU-wide voluntary agreement was realised.

Ingredients for Successful Policy

While only certain policies can specifically target CO_2 emissions, almost all transport (and many non-transport) policies affect CO_2 emissions from transport in one way or another. Therefore efforts to reduce CO_2 emissions should be embedded in overall transport reform in order to exploit synergies between transportation, environment and carbon concerns. Dealing with these problems almost always also restrains CO_2 emissions, hence the importance of exploiting the CO_2-transport policy link.

CO_2 will probably continue to be the "tail on the dog" as long as estimates of its external cost to society, and public concern about climate change, are less than for many other problems arising with transportation, such as air pollution, safety, congestion and sprawl. Already high fuel prices in many countries reduce the strength of arguments that costs associated with CO_2 emissions have not yet been internalised.

Four elements appear necessary for successful carbon restraint within the context of transport reform:

- Political consensus to make CO_2 reductions an important part of transport policy planning.
- Development of policy packages that have synergistic effects, for example that discourage automobile driving while investing heavily in programmes to improve alternative modes of travel.
- Careful assessment of the likely impacts of policies, and avoidance of overly optimistic projections.

- Successful implementation of the packages, with resources provided for on-going enforcement and reinforcement of policies.

In the short run, travel reductions and mode switching are the obvious places to turn for GHG reductions. But they are tough political choices. In the longer run, most countries believe that much larger CO_2 reductions will be made possible by the introduction of new technologies, such as "next generation" engines. But, as efficiency technologies do penetrate the market, they lower the fuel costs of travel, and countries may need to use pricing strategies to reduce travel rebound effects. This highlights the need for well-designed policy packages that work together to avoid sending mixed signals.

A critical job for governments is to convince citizens of the need for action, and educate people about how specific changes in transport can reduce CO_2 emissions. Without strong public support and understanding, there will be little political will to enact policies. Under such conditions, low-cost actions, even those that provide clear benefits beyond CO_2 reductions, can be blocked by individual stakeholders who may stand to lose. Governments also need to overcome "traditions" that no longer make sense: company car policies; very high fixed (but low variable) taxation; in a few countries, traditions of generally cheap fuel. Governments need to become more entrepreneurial: the presence of good transit or walking – biking facilities does not necessarily translate into high usage; the public has to be convinced that the benefits that these modes offer – lower traffic congestion, cleaner air, and lower CO_2 emissions – are worthwhile, and that they will be attainable only if people are willing to make changes to their life styles.

Policies should also recognise the importance of the starting point. Countries with a strong tradition of tight urban planning, good transit infrastructure, and non-motorised transport (like Denmark and the Netherlands) or a high share of low-carbon freight modes (like the US and Sweden) may have a hard time increasing the role of these factors. Conversely, countries with high levels of per-capita car travel and relatively large cars (again the US and Sweden), or high shares of energy-intensive trucking (Denmark, Germany and the UK) may have more opportunities to achieve carbon savings than those in the opposite situation.

The Outlook

It is difficult (and outside the scope of this study) to estimate the actual impact of individual policies on CO_2 emissions. Before-the-fact estimates of how individual policies will work are plentiful. For example, the IEA estimates that if the voluntary agreement in Europe works smoothly, emissions from cars will be 8 percent below what they would otherwise be by 2010 and 15 percent below by 2020. Country estimates of the potential impacts of policies and policy packages often indicate that they are expected to provide significant reductions in transport CO_2 emissions. However, the trend in transport CO_2 in all of these countries remains upward, with little obvious bending. While this is often discouraging, it does not necessarily mean that the policies are not "working." Underlying forces, such as growth in incomes, play an important role in shaping transport emission trends, and these forces can overwhelm the reductions derived from policy measures, especially when economic growth is stronger than expected. After the fact, it may appear as though the policies have not worked. In most cases, at least in the 1990ies, it is more likely that their benefits were simply overwhelmed by rising incomes and demand for mobility.

While recent experiences have shown how difficult it is to reduce CO_2 in transport, there is still considerable potential and cause for hope. There are many possibilities for CO_2 reductions that have not been fully explored by each country. A separate IEA report describing many potential new policies will be published in early 2001 to assist this effort. For example, rapid increases in the use of emerging information technologies (both to substitute for travel and to assist in making travel more efficient) have occurred in the past five years. This trend is still young. Governments will have an important role to play in seeing that these technologies are used for CO_2 reductions to the maximum extent possible. Governments will also play key roles in promoting the adoption of other new efficiency technologies as they become available.

REFERENCES

ECN/RIVM, 1999. *De Uitvoeringsnota Klimmatbeleid Doorgelicht, Een analyse op basis van het Optiedocument* [Climate Policy Implementation Plan, an analysis based on the Option Document]. Energieonderzoek Centrum Nederland and Rijkinstituut voor Volksgezondheid en Milieuhygiene (in Dutch).

Energy Information Administration (EIA), 1998. *Impacts of the Kyoto Protocol on US Energy Markets and Economic Activity*. Report # SR/OIAF/98-03.

ENS, 2000. *Teknologikatalog over Tekniske Muligheder for Energibesparelser i Transportsektoren* [Technology Catalogue over Technical Possibilities for Energy-savings in the Transport Sector]. March 2000. Copenhagen: Danish Energy Agency (in Danish).

European Conference of Ministers of Transport (ECMT), 1998. *Efficient Transport in Europe: Policies in Internalization of External Costs*. OECD/ECMT, Paris.

IEA, 1997. *Transport, Energy and Climate Change*. OECD/IEA, Paris.

IEA, 1997. *IEA Welcomes Kyoto Commitments*. Press release IEA/PRESS(97)38, 11 December 1997, Paris.

IEA, 1997. *Indicators of Energy Use and Efficiency*. OECD/IEA, Paris.

IEA, 1999. *CO_2 Emissions from Fuel Combustion: A new basis for Comparing Emissions of a major Greenhouse Gas*. OECD/IEA, Paris.

IEA/AFIS, 1999. *Automotive Fuels for the Future, the Search for Alternatives*. OECD/IEA, Paris.

SIKA, 1999. *Inriktningsplanerning och Koldioxidpolitiken* [Long-term Planning and CO_2 Policies]. ASEK Report. Stockholm. Sept. 1999 (in Swedish).

United Nations Framework Convention on Climate Change. www.unfccc.int

VROM, 1999. *Uitvoeringsnota Klimmatbeleid, Deel I: Binnenlandse Maatregelen* [Climate Policy Implementation Plan, part I: Domestic

Measures]. Ministerie van Volkshuisvesting, Ruimtelijke Ordening en Milieubeheer (in Dutch).

DENMARK

COWIconsult AS, 1994. *Tax Provisions in the Transport Sector. Vol. 1: Internalisation and the External Costs of Transport.* For Danish Ministry of Energy, EFP-91. Lyngby, Denmark: COWI.

COWIconsult AS, 1995a. *Subsidiering av Godstransport.* [Subsidies for Goods Transport] Notat Nr. 95.01. Copenhagen, Denmark: Transport raadet (in Danish).

COWIconsult AS, 1995b. *Analys av Bilafter 2005.* [Analysis of Automobile Fees 2005] Notat Nr. 95.05. Copenhagen, Denmark: Transport raadet (in Danish).

COWIconsult AS, 1996. *CO_2 Reduktioner I Transportsektorn. Arbeidspapir 1-7.* [CO_2 Reductions in Transport. Working Papers 1-7]. Copenhagen (in Danish).

Det Oekonomiske Raad, 1996. *Transport.* In Dansk Oekonomi, Foraar 1996. Copenhagen, Denmark: Det Oekenomiske 1996 (in Danish).

Energiministeriet (Ministry of Energy), 1990. *Energi 2000 Handlingsplan for en Bæredygtig Udvikling* [Energy 2000. Policy Plan for Sustainable Development]. Copenhagen: Min. Of Energy (in Danish).

Faerdselstyrsen (Danish Transport Authority), 2000. *Hvor langt Pa litreen – Personbiler 2000* [How Far on a Litre. Passenger Cars 2000]. Copenhagen: Faerdselstyrlsen (in Danish).

Kristensen, N.B., 1997. *Cost-Benefit Analysis of Policy Measures in the Transport Sector.* In Larsen, H. and Soendergaard, J., eds. Transport, Energy and Environment. Proceedings of the Danish Association for Energy Economics, Marienlyst, Denmark, Copenhagen.

OECD, 1995. *Motor Vehicle Pollution: Reduction Strategies beyond 2000.* OECD, Paris.

Schipper, L.J. and Eriksson, G., 1995. *Taxation Policies affecting Automobile Characteristics and Use in Western Europe, Japan and the*

United States. In Proceedings of 1993 Workshop "Sustainable Transportation", Washington D.C.: American Council for an Energy-Efficient Economy.

Telle, Martin, 2000. Private Communication.

Trafikministeriet (Ministry of Transport), 1990. *Regeringstranporthandlingsplan for Miljoe og Udvikling*. [Government Action Plan for Environment and Development]. Copenhagen: Ministry of Transport (in Danish).

Trafikministeriet, 1993. *Trafik 2005: Problemstillinger, Maal, og Strategier*. [Transport in 2005: Problems, Goals, and Strategies]. Copenhagen: Ministry of Transport (in Danish).

Trafikministeriet, 1995a. *Regeringens Handlingsplan for Reduktion af Transport-Sektorns CO_2 Utslip* [Government Action Plan for Reduction of the Transport Sector's CO_2 Emissions]. Copenhagen: Ministry of Transport (in Danish).

Trafikministeriet, 1995b. *CO_2, Energy, og Transporter. En Debat Oplaegg* [Energy, Transport, and CO_2: Debate Background]. Copenhagen: Ministry of Transport (in Danish).

Trafikministeriet, 1997. *CO_2-reduktioner i Transportsektoren – Hovedrapport* [CO_2 Reductions in Transport: Main Report]. Copenhagen: Ministry of Transport (in Danish).

Trafikministeriet, 1998. *Trafikredegoerelse 1997* [Transport Policy Review 1997]. Copenhagen: Ministry of Transport (in Danish).

Trafikministeriet, 2000. *Begraensning af Transportsektorns CO_2-udslip* [Limiting CO_2 Emissions in the Transport Sector]. Copenhagen: Ministry of Transport (in Danish).

Transportraadet, 1995. *Skrotningspraemien. Effekter for Miljoe og Bilpark* [Scrappage Incentive Impacts on Environment and Car Stock]. Rapport nr 95-04. Copenhagen: Transportraadet (in Danish).

GERMANY

BMU, 1994a. *The Federal Government's Decision of 29 September 1994 on Reducing Emissions of CO_2, and Emissions of other Greenhouse Gases in the Federal Republic of Germany On the Basis of the Third Report of the "CO_2-reduction" Interministerial Working Group* (IWG "CO_2-Reduction"). Bonn: Bundesministerium für Umwelt.

BMU, 1994b. *First Report of the Government of the Federal Republic of Germany Pursuant to the United Nations Framework Convention on Climate Change.* Bonn: BMU.

BMU, 1997a. *Beschluss der Bundesregierung zum Klimaschutzprogramm der Bundesrepublik Deutschland auf der Basis des Vierten Berichts der Interministeriellen Arbeitsgruppe "CO_2-Reduktions" (IMA "CO_2-Reduktion")* [Decision of the federal government with regard to the climate change action plan of the Federal Republic of Germany on the basis of the Fourth report of the interministerial working group "CO_2-reductions"]. Berlin: BMU (in German).

BMU, 1997b. *Second Report of the Government of the Federal Republic of Germany Pursuant to the United Nations Framework Convention on Climate Change.* Berlin: BMU.

BMV, 1992. *Bundesverkehrswegeplan 1992.* Bonn: Bundesministerium für Verkehr [Federal Infrastructure Plan 1992]. Bonn (in German).

BMVBW, 1998. *Aktuelle Themen: Verkehrsprojekte Deutsche Einheit-Verkehrs Infrastruktur Investitionen des Bundes in den neuen Bundesländern* [news update: German Reunification transport projects – transport infrastructure investments of the federal government in the new Länder]. Berlin: Bundesministerium fuer Verkehr, Bau- und Wohnungswesen (in German).
www.bmv.de/vde.htm

BMVBW, 1999. *Investitionsprogramm fuer den Ausbau der Bundesschienenwege, Bundesfernstrassen und Bundeswasserstrassen in den Jahren 1999 bis 2002* [Federal Ministry for Transport, Construction and Housing: Investmentprogramme for the Expansion of federal railways, federal roads and federal waterways from 1999-2002]. Berlin: BMVBW (in German).

BMVBW, 2000. *Kraftstoffe der Zukunft: Klimmt und Unternehmensvertreter wollen Strategie fue Markteinfuehrung neuer Kraftstoffe entwickeln* [Automotive Fuels for the Future: Minister Klimmt and industry representatives want to develop a strategy for market introduction]. Pressemitteilung Nr 9/2000, Berlin, 19. Januar 2000 (in German).

Bundestag 1996: *Entwurf eines Gesetzes zuer staerkeren Beruecksightigung der Schadstoffemissionen bei der Besteuerung von Personenkraftwagen. Document 13/4918, Deutscher Bundestag, 13 Wahlperiode, Bonn, Germany* [Draft Law to better take into account of pollutant emissions in the taxation of passenger vehicles] (in German).

DIW, 1997. *KFZ-Steuerreform: Nur geringe Umweltentlastung zu erwarten* [Vehicle tax reform: only small environmental benefits to be expected]. Deutsches Institut fuer Wirtschaftsforschung Wochenbericht 35/97, 28 August 1997, 64. Jahrgang, Berlin (in German).

DIW, 2000a. *CO_2 Emissionen im Jahre 1999: Rueckgang nicht ueberschaetzen* [CO_2 emission in 1999: reduction should not to be overestimated]. DIW Wochenbericht, 6/2000, Berlin (in German).

DIW, 2000b. *Die Deutsche Automobil Industrie im internationalen Wettbewerb* [The German automotive industry and international competition]. DIW Wochenbericht, 12/2000, Berlin (in German).

DIW, several years. *Verkehr in Zahlen* [Transport in figures], several editions. DIW, Berlin (in German).

Enquête Commission "Protecting the Earth's Atmosphere" of the German Bundestag (Ed.), 1994. *Mobility and Climate – Developing Environmentally Sound Transport Concepts*. Economica, Bonn.

Prognos/EWI, 2000. *Energiereport III – Die laengerfristige Entwicklung der Energiemaerkte im Zeichen von Umwelt und Wettbewerb* [Longer-term development of the energy markets under the constraint of environmental aspects and competition]. Verlag Schaefer-Poeschel, Germany (in German).

UBA, 1997. *Politikszenarien fuer den Klimaschutz* [Policy scenarios for Climate protections]. Berlin: Umweltbundesamt (ed) (in German).

VDA, 1995. *Freiwillige Zusage zur Kraftstoffverbrauchsminderung (22/3 1995)* [Voluntary Agreement for Reduction of Fuel Use]. Verband Der Automobileindustrie EV, Frankfurt am Main (in German).

THE NETHERLANDS

ECN/RIVM, 1998. *Optiedocument voor Emissiereductie van Boeikasgassen, Inventarisatie in het Kader van de Uitvoeringsnota Klimmatbeleid* [Option document for emissions reduction of greenhouse gasses, Inventory in the framework of the climate policy implementation plan]. Energieonderzoek Centrum Nederland and Rijkinstituut voor Volksgezondheid en Milieuhygiene, Den Haag/Bilthoven (in Dutch).

ECN/RIVM, 1999. *De Uitvoeringsnota Klimmatbeleid Doorgelicht, Een analyse op basis van het Optiedocument* [Examination of the Climate Policy Implementation Plan, an analysis based on the Option Document]. ECM/RIVM, Den Haag/Bilthoven (in Dutch).

Novem, 1996. *Econometer and Cruise Control, report of Field Trials*. Nederlandse onderneming voor Energie en Milieu. Novem, Sittard.

RIVM, 1993a. Nationale Milieuverkenning 3 [Third National Environmental Outlook]. Rijkinstituut voor volksgezondheid en milieuhygiene (in Dutch).

RIVM, 1993b. *Verkeer en Vervoer in de Nationale Milieuverkenning 3 en de SVV-Verkenning 1993* [Traffic and Transport Sector in the Third National Environmental Outlook and the Transport Structure Plan 1993], RIVM rapport nr 251701014, Bilthoven (in Dutch).

RIVM, 1996. *Transport Scenarios for the Netherlands for 2030: a description of the scenarios for the OECD project "Environmental Sustainable Transport."* RIVM, Bilthoven.

RIVM, 1997. *Effecten van Prijbeleid op Verkeer en Vervoer* [Effects of pricing policy on traffic and transport]. RIVM, Bilthoven (in Dutch).

RIVM, 1998. *Verkeer en Vervoer in de Nationale Milieuverkenning 4* [Traffic and Transport Sector in the Fourth National Environmental Outlook], RIVM rapport nr 251701014, Bilthoven (in Dutch).

V&W, 1990. *Tweede Structuurschema Verkeer en Vervoer, deel d, regeringsbeslissing* [Second Transport Structure Plan, Part D: Government Decision]. Den Haag: Ministerie van Verkeer en Waterstaat.

V&W, 1993. *SVV-Verkenning, Analyses en prognoses* [Second Transport Structure Plan, Analyses and Projections]. Den Haag: V&W (in Dutch).

V&W, 1996. *Beleidseffectmeting Verkeer en Vervoer, Beleidseffectrapportage 1995* [Measuring the Policy Effects in Traffic and Transport]. Den Haag: V&W (in Dutch).

V&W, 1996. *Transport in Balans*. Den Haag: V&W (in Dutch).

Van Wee, B. and Annema, J. A., 1999. *Transport and the Greenhouse Effect: The Role of Research in Kyoto Related Climate Policy in the Netherlands*. Nectar Conference, Delft October 1999.

VROM, 1993. *Vierde nota over de ruimtelijke ordening* [Fourth Report on Physical Planning]. Den Haag: Ministerie van Volkshuisvesting, Ruimtelijke Ordening en Milieu (in Dutch).

VROM, 1994. *The Netherlands' National Environmental Policy Plan 2*. Den Haag: VROM.

VROM, 1996. *The Second Netherlands' Memorandum on Climate Change*. Den Haag: VROM.

VROM, 1996. *Samen Werken aan Beriekbaarheid* [Co-operating on accessibility]. Den Haag: VROM (in Dutch).

VROM, 1999. *Uitvoeringsnota Klimmatbeleid, Deel I: Binnenlandse Maatregelen* [Climate Policy Implementation Plan, part I: Domestic Measures]. Den Haag: VROM (in Dutch).

SWEDEN

Association des Constructueurs Européens d'Automobiles, ACEA 1999. *Motor Vehicle Taxation in Europe, 1998-1999 edition*. Brussels. April 1999.

Edwards, H., 1999 (VTI 1999). *Utveckling av Transportsektorns Utslaepp av CO_2 fraan 1990 till 2010* [Development of Transportation CO_2 Emissions from 1990 to 2010]. Linkoeping: VTI (also published by SIKA) (in Swedish).

Government of Sweden, 1998. *Transport-politik foer en Haallbar Utveckling* [Transport Policy for Sustainable Development]. 1997/98:56 Stockholm: Riksdagens Tryckeri (in Swedish, English summary obtainable from Communications Department).

Jansson, J.O. and Wall, Rickard, 1994. *Bensinskattefoeraendringars Effekter.* [Effects of Changing Gasoline Tax]. Ds 1994:55. Stockholm: Dept. of Finance (in Swedish).

Johansson, O. and Schipper, L.J., 1995. *Measuring Long-Run Automobile Fuel Demand and Policy Implications: Separate Estimations of Vehicle Stock, Mean Intensity, and Distance Driven per Car per Year.* Proc. of The International Association for Energy Economics, Washington, DC 1995. Cleveland, OH: IAEE. See also Journal of Transport Economics and Policy. Sept. 1997.

KOMKOM, 1996. *Ny kurs I trafikpolitik. Delbetaenkande av Kommunikationskommitten om Beskattning av Vaegtrafiken* [A New Course in Transport Policy: Interim Report of the Communications Committee on Taxation of Road Traffic]. SOU 1996: 165. Kommunikationskommittén, Stockholm, Sweden: Fritzes (in Swedish).

KOMKOM, 1997. *Slutbetänkande av Kommunikationskommittén, 1997, Ny kurs i trafikpolitiken* (Final Considerations of the Communications Committee: A New Course in Transport Politics). KOMKOM, Stockholm, Sweden: Fritzes (in Swedish).

MATS, 1996. *På väg mot ett Miljöanpassat Transportsystem* [Towards an Environmentally Sound Transportation System]. Naturvårdsverket Förlag rapport 4636, Stockholm, Sweden (in Swedish).

SIKA, 1999. *Inriktningsplanerning och Koldioxidpolitiken* [Infrastructure Planning and CO_2 Policies] ASEK Report. SIKA, Stockholm: Sept 1999 (in Swedish).

TOK, 1994. *Trafiken och koldioxiden: Principer foer att minska trafikens kodioxidutslaepp* [Traffic and Carbon Dioxide: Principles for Reducing Emissions of Carbon Dioxide from Transport]. SOU 1994:91. Trafik och Klimatkommitteen, Stockholm, Sweden.

TOK, 1995. *Klimatfoeraendringar i Trafikpolitiken* [Climate Changes in Transport Policy]. SOU 1995:64. Stockholm, Sweden: Fritzes (in Swedish).

Trafikbeskattutredning, 1997. *Bilen, Miljoen, och Saekerheten* [The Car, the Environment, and Safety]. Trafikbeskattutredning, Stockholm: Dept. of Finance.

Volvo AB. Press release of Sept. 22, 1996 proposing lower fuel consumption.

Wallman S., (Volvo Car Corporation), 1996. Private Communication.

Westöö E., (Swedish Nat. Road Admin), June 2000. Private Communication.

THE UNITED KINGDOM

DETR, 1996. *Transport Green paper "Transport – the Way Forward"*. Department of the Environment, Transport and the Regions, London.

DETR, 1997. *Guide to Green Transport Plans, Advice to Government Departments*. DETR, London.

DETR, 1998a. *A new Deal for Transport: Better for Everyone (The Government's White Paper on the Future of Transport)*. DETR, London, July 1998.
www.detr.gov.uk/itwp/paper/index.htm

DETR, 1998b. *Climate Change Programme: Consultation Paper*. DETR, London, 1998.
www.environment.detr.gov.uk/consult/climatechange/index.htm

DETR, 1999a. *Environmental Impact of Road Vehicles in Use "Air Quality, Climate Change and Noise Pollution"*. DETR, London, July 1999.

DETR, 1999b. *Driving the Agenda, the First Report of the Cleaner Vehicle Task Force*. DETR, London, July 1999.

DETR, 2000a. *New Car Fuel Consumption and Emissions Figures*. DETR, London, January 2000. www.roads.detr.gov.uk/vehicle/fuelcon/index.htm

DETR, 2000b. *Tackling Congestion and Pollution, the Government first report under the Road Traffic Reduction (National targets) Act 1998*. DETR, London, January 2000.

HM/DETR, 2000a. *Budget sets Britain on road to better transport and environment*. HM Treasury/ Department of the Environment, Transport and the Regions Press Release, LONDON, 21 march 2000.

HM/DETR, 2000b. *Protecting the Environment; Reform of Company Cars Taxation*. HM Treasury/DETR Press Release, LONDON, 21 march 2000.

National Economic Research Associates (NERA), 1999. *A Framework for Assessing Studies of the Impact of Transport Infrastructure Projects on Economic Activity*. Report to the Standing Advisory Committee on Trunk Road Assessment (SACTRA), DETR, November 1999, London.

RAC, 1995. *Car Dependence*. Report for the RAC Foundation for Motoring and the Environment by ESRC Transport Studies Unit, University of Oxford with RDC Inc, San Francisco, 1995. ISBN 0 86211355 5.

SACTRA, 1994. *Trunk Roads and the Generation of Traffic*. Standing Advisory Committee on Trunk Road Assessment HMSO, London, 1994. www.detr.gov.uk/roads/roadnetwork/heta/hetasact.htm

The Stationary Office (SO), 1997. *Transport and the Environment, Developments since 1994*. Royal Commission on Environment Pollution, Twentieth Report, 1997.

Transport, 2000. *Changing Journeys to Work, an Employers Guide to Green Commuter Plans*. Supported by London first, 1998.

EUROPEAN UNION

Association des Constructueurs Européens d'Automobiles (ACEA), 1998. *ACEA Commitment on CO_2 Emission Reductions from New Passenger Cars in the Framework of an Environmental Agreement between the European Commission and ACEA*. Brussels.

Association des Constructueurs Européens d'Automobiles (ACEA). www.acea.be

Commission Task Force Car of Tomorrow, 1996. *Detailed Performance Targets and Research, Technological Development and Demonstration Priorities*. Draft, 6 June 1996.

European Commission, 1992. *The Future Development of the Transport Policy*. COM(92)494.

European Commission, 1994. *For a European Union Energy Policy*. Green paper, COM(94)659 final, January 1995.

European Commission, 1996a. *A Strategy for Revitalising the Community's Railways*. White Paper, COM(96)421 final, 30.07.1996.

European Commission, 1996c. *Toward Fair and Efficient Pricing*. White Paper, 1996.

European Commission, 1996d. *DGs for Industry, Energy and Environment*. The European Auto-Oil Programme, XI/361/96, 1996.

European Commission, 1997a. *Communication from the Commission to the Parliament, the Economic and Social Committee and the Committee of the Regions, Climate Change – The EU Approach for Kyoto*. COM(97)481 final, 1/10/97.

European Commission, 1997b. *Proposal for a Community Framework for the Taxation of Energy Products*. COM(97)30.

European Commission, 1997c. *Intermodality and Intermodal Freight Transport in the European Union – a systems approach to freight transport*. Strategies to enhance efficiency, services and sustainability. COM(97)243.

European Commission, 1997d. *Energy for the Future: Renewable Sources on Energy*. White Paper, COM(97)599 final.

European Commission, 1997e. *Car of Tomorrow – Summary of RTD Projects*.

European Commission, 1998a. *Climate Change – Towards an EU post-Kyoto Strategy*. COM(98)353.

European Commission, 1998b. *Communication from the Commission to the Council, the European Parliament, the Economic and Social Committee and the Committee of the Regions: On Transport and CO_2 – Developing a Community Approach*.

European Commission, 1998c. *Proposal for a Council Decision: Establishing a scheme to monitor the average specific emissions of carbon dioxide from new passenger cars*. COM(1998)348 final, Brussels 12.06.1998.

European Commission, 1998d. *Communication from the Commission to the Council and the European Parliament: Implementing the Community Strategy to Reduce CO_2 Emissions from Cars: An Environmental Agreement with the European Automobile Industry*. COM(1998)495 final, Brussels 29.07.1998.

European Commission, 1999a. *Commissions Communication to the Council and the Parliament: Preparing for Implementation of the Kyoto Protocol*. COM(1999)230, 19.05.1999.

European Commission, 1999b. *European parliament and Council Directive 1999/62/EC of 15 June 1999 on the charging of heavy goods vehicles for the use of certain infrastructures*.

European Commission, 1999c. *Explanatory memorandum to the proposals for national emission ceilings*. COM(1999)125 final, 09.06.1999.

European Commission, 2000a. *Commissions Communication to the Council and the European Parliament: EU policies and measures to reduce greenhouses gas emissions: Towards a European Climate Change Programme (ECCP)*. COM(2000)88, 08.03.2000.

European Commission, Directorate General for Energy and Transport in co-operation with Eurostat, 2000b. *EU Transport in Figures*. Statistical Pocket Book, January 2000.

European Commission,1996b. *The Citizens' Network: Fulfilling the potential of public passenger transport in Europe*. Green Paper, January 1996.

EVA/NOVEM, 1999. Brochure. *The Relative Fuel Economy Label.*

Fulton, Lewis, 2000. *Voluntary Agreements as Drivers of Technological Change in the Transport Sector*. OECD/IEA Project for the Annex 1 Expert Group on the UNFCCC, OECD, forthcoming.

THE UNITED STATES

API, 2000. *American Gasoline Institute website discussion of Oil Industry Position on Climate Change.*
www.api.org/globalclimate/apipos.htm

APP, 1998. *The Department of Transportation and Related Agencies Appropriations Act for Fiscal Year 1998*. Pub. L. 105-66, US Congress.

BP Amoco, 2000. *BP Amoco Statement on the Company's Commitment to reducing CO_2 Emissions.*
www.bpamoco.com/climatechange

CARB, 2000. *Low Emission Vehicle Program. California Air Resources Board.*
www.arb.ca.gov/msprog/levprog/levprog.htm

CBS, 2000.
cbs.marketwatch.com/archive/20000330/news/current/car.htx

CVC, 2000. *Coalition for Vehicle Choice.*
www.vehiclechoice.org/main.html

Dialogue, 1995. *President's Advisory Committee to Develop Policies for Reducing Greenhouse Gas Emissions from Personal Motor Vehicles.* Majority Final Report.

DOE, 1995. *Assessment of Cost and Benefits of Flexible and Alternative Fuel Use in the US Transportation Sector, Technical Report 14: Market*

Potential and Impacts of Alternative Fuel Use in Light-duty Vehicles: a 2000/2010 Analysis. Office of Policy, US Department of Energy.

DOE, 1999a. US Department of Energy Web site. www.ott.doe.gov/programs.shtml\#pngv

DOE, 1999b. US Department of Energy Web site. www.ott.doe.gov/oaat/pngv.html

EPA, 1999. *Light-Duty Automotive Technology and Fuel Economy Trends Through 1999*. US Environmental Protection Agency, Advanced Technology Support Division, Office of Mobile Sources.

EPA, 1995. Environmental Impacts of Removing National Speed Limit Requirements, memorandum. US Environmental Protection Agency. www.epa.gov/oms/reports/envspmem.htm

FHWA, 1998. *Transportation Equity Act for the 21st Century: A Summary*. US Department of Transportation: Federal Highway Administration. www.fhwa.dot.gov/tea21/sumcov.htm

Fulton et al., 2000. *A Statistical Analysis of Induced Travel Effects in the US Mid-Atlantic Region*. Journal of Transportation and Statistics, volume 3, number 1, April 2000.

Gallup, 1997. *Gallup News Service, "Public Concerned, Not Alarmed About Global Warming."* www.gallup.com/poll/releases/pr971202.asp

IEA, 1999. *Automotive Fuels for the Future: The Search for Alternatives*. OECD/IEA, Paris.

ISTEA, 1991. *Description of ISTEA and its programs*. osdbuweb.dot.gov/business/legislation/istea.html

Lave, Charles, 1992. *Cars and Demographics*. Access Magazine, University of California Transportation Center, issue #1.

NHTSA (National Highway Traffic Safety Administration), 1998. *Twenty-Third Annual Report to Congress*. Calendar Year 1998, US Dept. of Transportation.

Nivolla, Pietro and Robert Crandall, 1995. *The Extra Mile: Rethinking Energy Policy for Automotive Transportation*. The Brookings Institute, Washington D.C.

ORNL (Oak Ridge National Laboratory), 1999. *Transportation Energy Data Book: Edition 19*. Centre for Transportation Analysis, ORNL-6958.

PNGV, 2000. *PNGV official website.*
www.ta.doc.gov/pngv/introduction/intro.htm

Poll, 1999. *CNN/Time Poll taken January 20-21, 1999. 1,024 adults surveyed nation-wide.*
www.pollingreport.com/enviro.htm

Portland, 1996. *The Comprehensive Plan, 1996* (latest revision). City of Portland, Oregon, Bureau of Planning, 1900 SW 4th Avenue, Suite 4100 Portland, OR 97201.
www.planning.ci.portland.or.us

TEA-21, 1998. *TEA/ISTEA News.*
osdbuweb.dot.gov/business/legislation/istea.html

US EPA,1996. *National Air Quality and Emissions Trends Report.* Office of Air and Radiation.
www.epa.gov/airprogm/oar/aqtrnd96/chapter1.pdf

Wiener, Ed, 1997. *Urban Transportation Planning in the United States: An Historical Overview.* Fifth Edition.
www.bts.gov/tmip/papers/history/utp/toc.htm

Order Form

OECD BONN OFFICE

c/o DVG mbh (OECD)
Birkenmaarstrasse 8
D-53340 Meckenheim, Germany
Tel: (+49-2225) 926 166
Fax: (+49-2225) 926 169
E-mail: oecd@dvg.dsb.net
Internet: www.oecd.org/bonn

OECD MEXICO CENTRE

Edificio INFOTEC
Av. Presidente Mazarik 526
Colonia: Polanco
C.P. 11560 - Mexico D.F.
Tel: (+52-5) 280 12 09
Fax: (+52-5) 280 04 80
E-mail: mexico.contact@oecd.org
Internet: www.rtn.net.mx/ocde

OECD CENTRES

Please send your order by mail, fax, or by e-mail to your nearest OECD Centre

OECD TOKYO CENTRE

Landic Akasaka Building
2-3-4 Akasaka, Minato-ku
Tokyo 107-0052, Japan
Tel: (+81-3) 3586 2016
Fax: (+81-3) 3584 7929
E-mail: center@oecdtokyo.org
Internet: www.oecdtokyo.org

OECD WASHINGTON CENTER

2001 L Street NW, Suite 650
Washington, D.C., 20036-4922, US
Tel: (+1-202) 785-6323
Toll-free number for orders:
(+1-800) 456-6323
Fax: (+1-202) 785-0350
E-mail: washington.contact@oecd.org
Internet: www.oecdwash.org

I would like to order the following publications

PUBLICATIONS	ISBN	QTY	PRICE*	TOTAL
☐ The Road from Kyoto – Current CO_2 and Transport Policies in the IEA	92-64-18561-5		$75	
☐ Energy Technology and Climate Change – A Call to Action	92-64-18563-1		$75	
☐ World Energy Outlook 2000	92-64-18513-5		$150	
☐ Automotive Fuels for the Future – The Search for Alternatives	92-64-16960-1		$100	
☐ Experience Curves for Energy Technology Policy	92-64-17650-0		$80	
☐ Energy Labels and Standards	92-64-17691-8		$100	
☐ The Link between Energy and Human Activity	92-64-15690-6		$16	
☐ Transport, Energy and Climate Change	92-64-15691-7		$26	
			TOTAL	

*Postage and packing fees will be added to each order.

DELIVERY DETAILS

Name ____ Organisation ____
Address ____

Country ____ Postcode ____
Telephone ____ Fax ____

PAYMENT DETAILS

☐ I enclose a cheque payable to IEA Publications for the sum of US$ ________ or FF ________
☐ Please debit my credit card (tick choice). ☐ Access/Mastercard ☐ Diners ☐ VISA ☐ AMEX

Card no: ____ Expiry date: ____	Signature:

IEA PUBLICATIONS, 9, rue de la Fédération, 75739 PARIS CEDEX 15
PRINTED IN FRANCE BY CHIRAT
(61 00 29 1P) ISBN 92-64-18561-5 - 2000